T0313165

Ecological Risk Assessment

Innovative Field and Laboratory Studies

Ecological Risk Assessment
Innovative Field and Laboratory Studies

Lawrence V. Tannenbaum

CRC Press
Taylor & Francis Group
Boca Raton London New York

CRC Press is an imprint of the
Taylor & Francis Group, an **informa** business

The opinions or assertions contained herein are the views of the author and are not to be construed as official or as reflecting the views of the Department of the Army or the Department of Defense.

CRC Press
Taylor & Francis Group
6000 Broken Sound Parkway NW, Suite 300
Boca Raton, FL 33487-2742

© 2018 by Taylor & Francis Group, LLC
CRC Press is an imprint of Taylor & Francis Group, an Informa business

No claim to original U.S. Government works

Printed on acid-free paper

International Standard Book Number-13: 978-1-4987-8617-1 (Hardback)

Library of Congress Cataloging-in-Publication Data

Names: Tannenbaum, Lawrence, author.
Title: Ecological risk assessment : innovative field and laboratory studies / Lawrence V. Tannenbaum.
Description: Boca Raton : Taylor & Francis, 2017.
Identifiers: LCCN 2017034217 | ISBN 9781498786171 (hardback)
Subjects: LCSH: Ecological risk assessment.
Classification: LCC QH541.15.R57 T36 2017 | DDC 333.71/4--dc23
LC record available at https://lccn.loc.gov/2017034217

Visit the Taylor & Francis Web site at
http://www.taylorandfrancis.com

and the CRC Press Web site at
http://www.crcpress.com

A wise man will hear, and will increase learning; and a man of understanding shall attain to wise counsels. (Proverbs 1:5)

೫ ೫ ೫ ೫ ೫ ೫ ೫ ೫ ೫ ೫ ೫ ೫ ೫ ೫ ೫ ೫ ೫ ೫

With unbounded gratefulness to the Creator of the Universe for His gift of life, and for bestowing me with such a wonderful family. I dedicate this book to

– My Aishes Chayil, Chava Esther,

– Our dear children Eliezer Moshe and his wife Hindy; Shifra Miriam and her husband Yaakov Boruch

– Our dear grandchildren Dovid Yaakov, Freida Leah, Yechezkel, and Malka Perel Tannenbaum; Shmuel, Sheftel Yekusiel, and Freida Leah Beren.

May we be able to continue to help preserve this wonderful land for future generations.

Contents

SECTION II   Laboratory Studies

SECTION III   Desktop Studies

Preface

Perhaps I am properly typecast as an "ERA practitioner." While I can't be sure that this term exists by virtue of someone having formally coined it, the term seems to work. The appellation seems fitting in my case, in light of the work duties I've assumed for more than 25 years. These include reviewing work plans for ecological risk assessments (ERAs), reviewing ERAs themselves as stand-alone documents or as they occur within remedial investigations, writing in-house ERAs, serving as a subject matter expert where ecological risk is at issue, teaching course units on ERA, and researching and publishing on novel aspects within the field. If ERA occupies considerable amounts of your time, involving any or all of the tasks I list, or still others, I formally welcome you into the camp and cordially invite you to work your way through this book. If you have a thirst for wanting the best ERA science to come about, I would say you are most definitely a kindred spirit, and my book is here to inspire and motivate you.

I think it's fair to say that this book is novel in design. Although I haven't looked into it, I don't have the sense that for other fields of science (e.g., geology, chemistry, metallurgy), professionals have published books describing useful studies to conduct. Indeed, the book's title could be *Ecological Risk Assessment: Worthwhile Studies You Might Not Have Thought About That Can Advance the Field*. Encouraging others in one's field to take a more active research role is one thing, but giving away personally cultured research ideas that can well seed the successes of others is another, and perhaps a bit odd. No, it's not that I'm so particularly altruistic. My free dissemination of ideas is explained just a few paragraphs further down.

This book is future-oriented in two ways. First, it is intended to advance the ERA field beyond where it presently and unimaginatively stands. It is also future-oriented because it discusses studies that have yet to happen. The book is a compendium of studies for which I feel certain that findings produced will lead to a much-improved understanding of our ERA lot. Further, with the proper dissemination of the data to be gathered through conducting the studies, that "much-improved understanding" could lead to a major overhaul of ERA process, something I believe we direly need.

The premises for the studies are sourced in status quo ERA and the rank and file's complacency with it. For me, there is not a business day that goes by where, on my ride home after work, I don't pinch myself, asking questions such as "Have I colleagues who really think that site ecological receptors at 80-year-old sites have a potential health risk looming?" "Could there honestly be utility to a toxicity test that involves deliberately removing the outer protective membrane of an aquatic animal's embryo?" "Do we really need to understand the ecotoxicity of explosives in soil and, in particular, for species we cannot observe in the wild?"

The studies described fall into three categories: field studies, laboratory studies, and those to be conducted in the office where one avails himself or herself to the vast stores of information residing in publications, on the Internet, etc., and then integrates information at the desktop level. A number of studies have both laboratory and field components to them. In those cases, their categorization is generally in line with the dominant study aspect. I have no doubt that the studies can provide the bases for master's theses and PhD dissertations. Also, the information to be gleaned from the studies can provide the substance of technical papers to be published in the peer-reviewed literature. There are other features of the described studies to which I can alert you. For starters, they are of the one-of-a-kind genre. The studies you will be reading about, then, are not the kind we would hope that various regulatory agencies would latch onto and incorporate in ERA guidance such that they become ERA fixtures. These are studies, though, that once completed, should make certain notions eminently clear—that some ERA tasks are fully unnecessary, select measures or computations are erroneous, many species have no place whatsoever in assessments... and the list goes on. And brace yourself; more than illustrating that ERA for conventional applications needs to be completely reworked, the information to flow from the completed studies has a good chance of informing us that there was never a need for an ERA process in the first place. The studies, too, should indicate what tasks ERA (or what ERA is to ultimately morph into) should be adopting. While the studies are not meant to be regularly applied, it would be ideal if, in the short-term, multiple parties invested their energies in playing them out. Here, in particular, I have the field studies in mind. We live in a world of diverse habitats, and to the extent that a key goal is to elucidate truisms and trends, there is a need to know whether or not, across a highly variable landscape, animals behave or respond similarly to stimuli. If they do not, we can at the very least identify when it is appropriate to include a certain receptor or to assess it in a particular way, and where not to do such. You will note in the study write-ups the frequent encouragement of multiple interests to come forward, and the repeating call for a consolidation of the findings of independently applied efforts. Importantly, while the study guidelines provided are intended to see researchers through to their successes, at the end of the day, each study is a straw-man proposal. Interested parties are free to modify them as they see fit, and this leads to a rather important point.

You will note that the studies presented have a formatted structure. A premise, usually two paragraphs long, frames the situation worthy of study, culminating with a brief statement on what the (hopefully) enthused scientist is being asked to do. The study guidelines, while mostly of distinct tasks to accomplish (e.g., assemble a list of sites with a certain feature; rig up a harness to comfortably restrain a bird), also provide notes that shape the study (e.g., that no lab animals are to be used). The outcomes discussion bears on seeking opportunities to disseminate study findings, always with the goal of having a bettered ERA process emerge. You are forewarned that, other than for the introductory chapter, there are no references supplied. This book is about the future; the thrust is not on documenting the problems that exist or studies that have been conducted in the past, but rather on what could be done to get beyond the problems and the past studies. In a very real sense, supplied references for phenomena that we know of would present as distractions. As mentioned above, it is the information the studies stand to unleash that I hope will seed the references of tomorrow that others seek.

These studies are there for the taking. While the author would personally like to be involved with them all, such is not a realistic view. Since I can't get to them, I can only hope that you, the reader, will rise to the occasion. If you yourself won't be able to get at them, perhaps you've colleagues whose time, schedules, and interests permit. While writing this book, I have pondered the question of whether or not (any of) the studies will ever happen. I compiled a mental list of reasons why ERA practitioners or students of environmental science might not engage with them. These span quite a range—from the studies being too unusual or too difficult to pull off, to parties being disinterested in making forays into them because (innovative as they truly are, I believe) they did not come up with the ideas themselves. I don't mind sharing with you that, while developing the book and anticipating that none might come forward to run the studies, I contemplated book titles along the lines of: *Ecological Risk Assessment: Useful Studies We Can Only Hope Will Be Conducted*. I didn't dwell on these negative thoughts for very long. I have the sense that even if the studies do not materialize, there's plenty of good that the book you hold can still accomplish. For the dedicated ERA practitioner, the book will stimulate thinking and cause individuals to shed timidity, empowering them to step up and challenge the status quo. Even with study outcomes unknown, I imagine, for example, an individual who would otherwise tacitly concur with a report describing a toxicity test now vocalizing, with confidence, that the test results don't necessarily provide anything worthwhile and perhaps, too, averring that the test should not have been run.

Hardest for me to write were the study sections titled "Study outcomes and applications thereof." No one has a crystal ball, and any scientist worth his salt knows that the data must speak for itself. Much as I tried to have the concluding sections present the ramifications of diametrically opposed outcomes, I see that I have tended to favor writing about what I anticipate the outcomes will be. Consider that the objective behind running the studies is to bring forward information to either say that certain ERA tasks are purposeful and grounded or that they are not (and maybe, too, to say whether the whole of ERA is purposeful and grounded or not). As for you, the reader, who will hopefully engage with some of the studies, I remind you of biology's first law: the animal is always right. There's nothing wrong with having an expectation of what a study's findings will show, but please do not let your hunches bias you in your efforts. If what lures you to conducting any of the studies should be the interest in proving me wrong, that's perfectly fine. Either way, you'll be constructively contributing to bettered ERA science.

A last note: please stay in touch, alerting me to what you've found. I eagerly await the findings that only you can bring forward.

Lawrence V. Tannenbaum
Baltimore, MD

Author

Lawrence Tannenbaum is a biologist/ecologist and a chemical exposure health risk assessor for human health and ecological concerns. He began his career in risk assessment when working in the Superfund Program of the U.S. Environmental Protection Agency in its Region II office. For more than 20 years he has been employed by the U.S. Army, working on a diverse range of risk assessment applications. His areas of high interest include documenting the recurrent absence of impacts in ecological species contacting conventional hazardous waste sites, accounting for that phenomenon, and in illustrating that conventional ecological risk assessment methods are ineffectual. To his credit, Mr. Tannenbaum is the sole inventor of a patented field-based method for assessing the health of chemically exposed mammals, the only method of its kind. He is a senior editor (health and ecological risk assessment) for the SETAC journal, *Integrated Environmental Assessment and Management*. He is the author of *Alternative Ecological Risk Assessment: An Innovative Approach to Understanding Ecological Assessments for Contaminated Sites* (Wiley-Blackwell, Sussex), presently housed in more than 500 university libraries in the United States and in numerous other countries. He has published more than 40 papers on varied risk assessment topics in peer-reviewed journals.

Mr. Tannenbaum holds bachelors and master's degrees in biology from the City College of the City University of New York and is a certified senior ecologist through the Ecological Society of America.

Introduction

The starting point for this book is, unfortunately, a negative one. It is the recognition that the discipline of ecological risk assessment (ERA) as it is applied to discrete contaminated sites, such as Superfund sites, is severely lacking and undeveloped. Unbiased professionals working in this field would be hard-pressed to argue the point. Consequently for this field's genuinely enthused and motivated practitioners who hopefully exist in respectable number, the current status of ERA must be truly disappointing. The facts are these: 1) Sophistication in assessing the health state of ecological receptors as they exist today at contaminated sites, or in forecasting the possible health impacts to accrue to them, is all but limited to the construction of ratios at the desktop. These either take the form of comparisons of chemical concentrations in environmental media with supposed safe or toxic effect levels (as tabular/look-up values), or computed diet-based hazard quotients (HQs). 2) There is an evident complacency with the ERA process in place, and there are no indications of a move afoot to bring about ERA reform that, minimally, would entail dispensing with HQs and the like. 3) While ecological and ecotoxicological research bearing on various ERA components proceeds at a steady pace, inherent design flaws in the research greatly lessen, or strip away entirely, the utility of the new information being brought forward. While this has been, without question, a negative ERA status assessment, there are positive steps we can take to get beyond it.

The starting point for embarking on truly utilitarian studies in support of ecological assessment work can only be the acknowledgment of the design flaws just mentioned and conceding, in full, that newly acquired information, novel as it may be, but stemming nevertheless from studies with inherent design flaws, is absolutely worthless for ERA. Even with problem-plagued studies, scientists are wont to find at least some saving-grace extractable element that they feel breaks new ground, or perhaps points in a new direction to be explored. Harsh as it may sound, this positive-spin perspective can be severely counterproductive, and certainly so for a field of endeavor that, for the time being, doesn't recognize how

1

weak and undeveloped it is. Rather than salvage something from well-intended studies that nevertheless fail to furnish useful information, we would be better off to dismiss the studies entirely, and to begin building anew. With this (admittedly somewhat ambiguous) preface in place, it is time to move onto a few concrete examples of studies that are well-intentioned in an ecological assessment context but that unfortunately bark up the proverbial wrong tree.

ERA concerns over the chemical exposures of avians recognize that opportunities for birds to succumb or, minimally, to accumulate what may be unhealthful chemical residues do not simply commence with feeding and other activities in post-hatch life. Realistically, maternal transfer to the egg occurs as it develops within the oviduct, reflecting the lifetime chemical exposures of the dam through parturition. It would, of course, be worthwhile to understand the attendant health risks (if not the all-out impacts) with which a bird is presented at the time of hatching, i.e., before it will invariably incur additional chemical exposures through dietary and other intakes occurring from that point on. This knowledge could presumably inform that developmental effects are a given in birds with a certain congenital body burden. Birds may be fated to display errant behaviors, to not achieve their full size, or to demonstrate developmental delays that will go on uncorrected. For a goodly number of ecotoxicologists, dosing fertilized eggs with chemicals via the cautious penetration of the egg shell with a hypodermic needle has become a study method of choice (Winter et al., 2013). Chicken (*Gallus domesticus*) and Japanese quail (*Coturnix japonica*) have become the predominant test species where the effects of xenobiotics introduced during embryonic development is the topic of study. In a short span of years, great attention has been given to perfecting dosing techniques so that mortality and deformities related to this can be segregated from effects that arise only from embryonic exposure to a test chemical, perhaps an endocrine disrupter (Ottinger et al., 2005). Thus, studies have proceeded that qualify and quantify egg-injection-related health effects in terms of the delivery site within the egg (e.g., the air cell, a pocket of air that is formed at an egg's blunt side as a result of cooling-induced contraction of egg contents after laying), and the timing of injection (i.e., embryonic age). With this basic introduction to a specific experimental design type, it is time to consider whether the design has merit, and particularly in that context where toxicity reference values (TRVs) are to be developed for use in desktop ERAs.

Better-quality egg-injection studies will consider multiple locations for chemical delivery, as perhaps the yolk and the albumin in addition to the air cell. Other experimental design variables will include dose and volume of the substance injected. Importantly, if we imagine a series of well-controlled studies that incorporated the variables mentioned here, and even additional ones, we would not be able to glean any useable information from this, and, in particular, a technically sound TRV could not be derived. We would do well to itemize the caveats of the study design as described to appreciate this, and to appreciate, too, how in the guise of a qualified technique and thoroughness of its implementation, grave inadequacies of the data to be generated can be overlooked. Acknowledging, then, that chemical delivery through the eggshell via needle and syringe never occurs in the real world fails to even scratch the surface of this discussion. Two far-reaching consequences of the egg-injection system must be specifically called out. First, while data may be lacking to illustrate it, it is quite unrealistic to posit that, in nature, the entire quantity of a maternal-source xenobiotic appears in an embryo in an instant, corresponding to the few seconds it takes to dispense a chemical dose from a syringe. It is equally unrealistic to posit that the entirety of a maternal-source xenobiotic or naturally occurring chemical (that will be giving rise to an excessive/unnatural tissue burden) will find itself localized to a singular region within an incubating egg, corresponding to the very place at which the tip of the hypodermic needle was positioned when the syringe plunger was squeezed. But before all of this, it is most unreasonable to suggest that the maternal-transfer—all of it—occurs in the nest, i.e., during the egg's incubation outside the mother's body. How could it be that of a dam's toxic load, as either a xenobiotic or an excess of a naturally occurring inorganic compound (e.g., copper), no portion of it is transferred to the egg in the oviduct either via the bloodstream before shell deposition, or via gas exchange after shell deposition? Whether fully recognized or not, egg-injection studies are considering that a toxin only first appears in an egg after it is laid. Is that the case, though?

The experimental egg-injection system raises more questions than it answers. There do not appear to be opportunities for an incubating egg in a nest to absorb a chemical load. The featherless brood patch has never been described in the literature as a rich and overflowing chemical source that generously infuses eggs through the porous shell. Soil-to-plant chemical transfer, in a general sense, is not an overriding phenomenon at contaminated sites, and thus the twigs and grasses used to fashion nests are unlikely to bear sizeable stores of site-bound contaminants to dose the eggs that might lie against these materials. While the specific egg compartment into which a chemical load is artificially delivered might well correlate with instances where developmental effects take hold, it seems unlikely that potentially harmful toxic stores that arise in nature are confined to a singular compartment. If chemicals of either category that concern us (i.e., of the type that don't belong in eggs altogether, or that could be expected to accumulate in eggs but not necessarily in alarming quantities) become stored in multiple or all egg compartments, of what utility are compartment-specific egg-injection studies? Consider that, while the experimental exogenous introduction of estrogen induces transformation of the left testicle into an ovotestis in males, and causes persisting Mullërian ducts and duct anomalies in males and females, respectively (Brunström et al., 2009), eggs in the real world are not exogenously exposed to estrogen.

The exogenously dosed bird egg needn't be the lone poster-child of eco-toxicological studies in serious need of repair. That is, there is no shortage of studies from which to pick, where a critique is in order for their not having contributed to an improved understanding of how an ecological receptor fares in the face of contamination, though such was the studies' intent. In fairness, though, long before any energies are expended on discrediting the hard work of others, we must recognize that egg-injection and other chemical exposure studies are not always conducted to assist ERAs for contaminated sites. Pure, as opposed to applied, toxicological research has its *bona fide* merits. And so, what has been discussed to this point could all be considered in the vein of "What would happen to a bird's hatching or its overall development if, all of a sudden, a chemical were to be administered to an incubating egg in a fashion far-removed from the uptake mechanisms that are at play for wild birds at contaminated settings?" The ever-curious scientific mind might wish to know if there is a critical window within the incubation phase where irreversible damage sets in. Critical-window studies can always have their rightful and constructive place and, where directed at ontological events or at refining concepts in organic evolution, can undoubtedly be highly informative. Coordinated research, for example, has pinpointed the specific day in a mouse's gestation where the delivery of a singular intoxicating dose of alcohol to the pregnant (and not previously exposed) mother will result in craniofacial deformities that translate into behavioral and other impacts to its young. Such information woven into an exercise aligning a mouse's 21-day gestation with man's 40-week gestation has helped to identify the corresponding point in a woman's pregnancy where limited alcohol consumption will lead to the complex and damaging neurological and psychiatric symptoms of fetal alcohol syndrome (Sulik et al., 1981). So, then, a critical-window chemical administration study, such as one involving bird egg injection, can be worthwhile. Our human antennae need to be deployed, though, where experimentally induced chemical exposures that fail to mimic those occurring in nature are showcased in the platform sessions and posters of ERA-oriented conferences and symposia. With many ERA practitioners already misled with regard to the applicability of the questionable-worth ecotoxicological information they come across, it is incumbent on us to lessen the opportunities for them to be further misled.

The fact remains that egg-injection studies *do* proceed for the purposes of crafting lowest observed adverse effect levels (LOAELs) and related toxicological benchmarks for use in desktop ERAs (Molina et al., 2006). With this in mind, it would be a profitable exercise to work backwards from these studies with the ultimate intent of arriving at the precise ecological assessment question(s) that need to be asked. From there, we'll be primed to design truly appropriate studies as needed, to tackle specific and relevant questions that can arise in ERA work. And so:

- Do we know of instances in nature where a site contaminant of concern appears in a singular location in an incubating egg? A literature review could easily tackle this first query. In a structured way, all those studies involving the collection of eggs in the wild, at or in the vicinity of a contaminated property, would first be assembled. The studies would be reviewed to see if chemical loading to the egg had been assessed. Chances are that such assessment work was not done, with researcher interest having yielded instead to collecting egg-size or clutch-size data in support of a comparative assessment with pristine areas in the same state or region as the contaminated property. Where a literature review of the type described here comes up dry, we learn that egg-injection studies are without the footing needed to assist with chemical exposure issues of birds in the wild and, in particular, the furnishing of TRVs.
- If, by some quirk, the contents of field-collected eggs should have been submitted to chemical analysis, tracking how the analysis was specifically carried out would be a critical next task. If the egg contents were spilled out and mixed or homogenized, all opportunity to document contaminant localization to a specific compartment will have obviously been lost, although the level of contamination in the shell as a discrete tissue could still be ascertained. (Whether or not there is a need to know a shell's chemical content is another matter. That would first be dependent on the knowledge that a substantial fraction of an egg's total chemical content sequesters to the shell, a phenomenon undoubtedly chemical-specific if it should be operative at all.) Even if eggshells should be a dominant chemical repository, such would only be meaningful in a food-chain context if the shells constituted a sufficiently large dietary percentage of an avivorous/ovivorous species that we aim to protect. If the literature review should be devoid of egg-compartmentalized chemical analysis information, as is likely to be the case, it would become

clear again in an overall sense, that *in ovo* (injection) dosing studies do not relate in a meaningful way to ERA needs.

- It is possible that, before eggs were cracked open to expel the contents, they were needled to remove air or specific tissues for chemical analysis, perhaps to remove a gas sample from the air cell. Where that was the case, air-cell injection studies (such as with perfluorooctane sulfonate [PFOS]; Molina et al., 2006) could potentially still have merit, and two contingencies would decide this. We would first have to know that chemicals sequester to the air cell. Assuming they did, for each chemical detected in the air cell we'd need to know the concentration that contravenes the health of the embryo and/or the hatched bird at some later time. It does not appear that either of these information types exist, leading us to the recurrent question of what utility lies in having artificially placed a chemical load to a specific region in the egg if correlative data doesn't exist for egg-compartment tissue concentrations and associated health effects. Where this two-step screening demonstrates absent study utility, there is an obligation upon the ecological assessor to critically observe how well-intentioned researchers manage nevertheless to declare gains. The for-the-moment "poster-child" study of air cell-injected PFOS in white leghorn chickens serves us well instructionally, and we begin with a review of the study's relevant elements.

1. Prior to incubation, the chicken eggs were injected with a range of PFOS doses, but there were no cases of significant difference in the number of embryo abnormalities between vehicle-dosed eggs and eggs of any of the PFOS-dosed egg treatment groups. While the immediate focus of this discussion is not on the required degree of demonstrated adversity in treated study groups that allows for the derivation of effect levels such as the LOAEL (Tannenbaum, 2014), one thing is clear. In terms of chemically induced morphological change to the embryonic liver, there had been no evidenced adversity.

2. The study authors nonetheless keyed into observed morphological changes in the embryonic livers of the PFOS-injected eggs, citing mild hepatocellular vacuolation and, at the same time, noting that this condition is not always injurious to the liver.

3. The study authors reported another observed morphological change in the embryonic livers of PFOS-dosed eggs, namely bile duct hyperplasia. At the same time, they noted this to be a common lesion, and they stated as well that it remains unclear whether this condition can progress into benign or malignant tumors. The authors further noted that the hyperplasia may remain static, regress over time, or progress to a more proliferative state, and also that no embryonic livers of PFOS-dosed eggs had evidenced the more proliferative state of cholangiofibrosis.

4. Based only on the morphological changes described in points 2 and 3, the study authors nevertheless proclaimed a LOAEL in the form of the PFOS dose delivered to the pre-incubation egg (of 1.0 µg/g egg, the next-to-lowest of the study's multiple doses). The study author's rationale for setting this LOAEL constitutes the true object of this chapter's attention. Although PFOS toxicity had not been shown, the authors remarkably contended that, should wild avians be as sensitive to PFOS as are chicken embryos, PFOS may have a detrimental effect on birds inhabiting contaminated areas. The rationale collapses further with the researchers' argument put forth: because the egg injection route of exposure had resulted in hepatic concentrations of PFOS similar to those measured in northern bobwhites and mallards in the wild—species more likely than chickens to be exposed naturally to PFOS, it would seem—the egg-injection method can be used as a rapid screen for the effects of other similar perfluorinated compounds.

The reader should be stupefied at the chemical-effect associations and extrapolations of the authors and study directors. Since when can *embryonic* liver concentrations be compared to liver concentrations of mature birds? How do we know that the white leghorn chicken can serve as a surrogate for mallard and northern bobwhite, and seemingly without a dose adjustment for different life stages? How can egg-injection data be compared to naturally occurring maternal deposition data? What basis is there for establishing a dose or concentration equivalency for an injected egg and an adult bird liver, i.e., that PFOS *in the egg* at 1 µg/g translates into 1 µg/g in the liver of a wild bird? How is it possible for the authors/study directors to claim they have assembled a working assessment method for perfluorinated compounds, recalling that an assimilated tissue concentration (of a xenobiotic or of any chemical type, for that matter) is not itself an "impact," and when no actual impacts have been identified?

What were the study directors researching? With the terminus of their efforts being a LOAEL, there can be no question that an inroad for ERA was being sought. Presumably, the story begins with the study directors having an appreciation that PFOS, a man-made fluorosurfactant with a global distribution, can harm birds that accumulate the compound in various tissues. Echoing that appreciation are the numerous efforts conducted over the last two decades or so, involving a variety of wildlife species (e.g., eagle, cormorant, polar bear, mink, dolphin, etc.), where PFOS levels have been measured in a range of tissues including plasma, liver, and muscle (Houde et al., 2006). While it is laudable to try to limit PFOS accumulations in birds so as to limit potential health and population impacts, such is not possible where an exceptionally stable compound with a world-wide distribution is concerned. Clearly there will be no efforts expended to reclaim PFOS from the environmental matrices where the compound today exists, and, thus, responsible actions from this point forward can only take the form of placing bans on the further manufacture and use of this potentially problematic persistent organic pollutant. With this understanding, the purpose served by a crafted LOAEL, assuming it is reliable, is in enabling man to identify when birds in the wild are at health risk because their eggs or whole bodies bear excessive levels of PFOS. The PFOS LOAEL espoused by the highlighted study's authors, then, is not a concentration to be used to indicate the degree to which the PFOS environmental load would need to be reduced such that birds would no longer be at risk, again because there will be no attempts made to reduce that load. Having crafted a LOAEL in the form of a standing tissue concentration in a specific organ (be it embryonic or adult-state) is misleading for the ERA field in other ways; conventionally, LOAELs and other toxicological benchmarks (e.g., LC50) are based on either internally or externally administered doses. Moreover the study's LOAEL, a screening tool, to be used to indicate when a bird is fated to develop illness or die, is highly impractical. Getting at a wild bird's liver PFOS concentration necessitates animal sacrifice. Perhaps too, then, the study directors haven't fully thought through their invention. Several wild birds are collected, euthanized, and have their livers harvested for PFOS analysis. If PFOS is not detected in the livers, or is detected at less than 1 µg/g, we would seemingly know that had the birds not been euthanized, they would have lived on without incident—at least as far any PFOS in the environment is concerned. Had PFOS liver concentrations of 1 µg/g or greater been detected, the study directors would have us believing that those birds were fated to develop pathological changes in the liver (if they hadn't already done so) that would, in turn, have led to death. Does it matter that a bird develops pathological changes to the liver, i.e., changes that we earlier saw are in the normal range? What's wrong with a bird that, upon examination, exhibits a changed liver? Will we conclude that those birds whose livers bore PFOS concentrations above 1 µg/g would have died prematurely had they not been euthanized?

There is no need to dwell any longer on the egg-injection ERA calamity, and soon enough the discussion will move onto a second example of an unfortunate and corrupted experimental design also intended to explore an exposure of concern not often addressed. For now, it is pivotal that the reader appreciate the foregoing analysis for what it is. In short, a toxicological approach was applied to an ecological risk topic. A closer look reveals that the toxicological effort was not only profoundly lacking, but it could not shed light on the question that spurred the effort on. Perhaps the most egregious error with the testing that was done was that the applied toxicology was to the exclusion of any integrated *ecological* awareness or science; in fact the word "ecology" does not appear even once in the highlighted study. Strides in the ERA field will not come about if the approach to study always takes the form of chemically dosing animals.

Before launching into one other example of failed ERA assistance, it would be extremely worthwhile to ask what could have been done to responsibly address the ecological concern that has been discussed in detail over the last few pages. Studies or queries that would target the ecological concern at its core would be the order of the day. Dare it be said, nowhere near the top of the list of possibilities to explore should there be planned laboratory studies involving the use of hypodermic needles and the like. Perhaps the best place to start is with an analysis that can discover if the ecological risk question itself even starts! We should first ask (and seek to reliably answer) the question of bird populations declining, and specifically in an overall sense, since it has been established that PFOS in the environment has a global reach. We could hone the question to a consideration of mallards and northern bobwhites, since these were species made out to be at risk, given the PFOS levels that had purportedly been said to match up with those of birds dosed in the lab. Critically, great attention should be directed at discrediting the original question, since examples of chemically imperiled birds and mammals are, for all intents and purposes, unknown.

If information suggesting population declines is lacking, concern over birds bearing toxicological insults from PFOS exposure will have been found to have lost a great deal of steam. If more than a few bird populations should be showing declines, we would still be far from having clinched anything scientifically; why would one's thoughts turn to PFOS as the culprit, recalling that there are hundreds of contaminants in the environment?

A next opportunity to peer into the world of well-intended but nevertheless scientifically compromised studies that some may latch onto when seeking ERA assistance concerns reptiles, and specifically lizards. A fair amount of studies have been published that examine the toxicological consequences of fence lizard (*Sceloporus sp.*) exposures to explosives (McFarland et al., 2008, 2009; Suski et al., 2009). These studies understand that expansive U.S. military installations are often home to sizeable land parcels bearing residual explosive concentrations, as well as numerous reptilian forms. The common laboratory-based experimental design of the studies begins with preparing dosing suspensions of an explosive in an appropriate inert vehicle (e.g., RDX in corn oil; 2,4-DNT in methyl-cellulose), with compound dissolution brought about through stirring a mixture sitting atop a heat source. Dosing by pipette to the rear of the mouth is accomplished by researchers working in tandem. One researcher coaxes the mouth open using either a guitar pick or by pulling down on the anterior end of the mandible with the nail of the index finger. While maintaining a consistent gentle downward grasp of the dewlap, the second researcher delivers the suspension via a hand-held pipette. Based on the review of differential responses in treated and control animals of 60-day dosing experiments, NOAELs and LOAELs for a diverse range of end-points are developed. These include hematological indices, growth rates, organ weights, sperm features, blood chemistries, and post-dose observation. Before embracing this relatively new brand of investigation that is so attentive to a neglected animal group (i.e., reptiles as a whole), a judicious review, not unlike that for the previous case considered, is in order. The intent of the review will be to concretize the point that the first step leading to the development of truly utilitarian studies to assist ecological assessments is a deliberate effort to critique and discredit existing studies. A key qualification is in order here before proceeding with such, however. The suggested deliberate critiquing (debugging is probably a more accurate term) and discrediting is specifically targeted for ERA, an applied science, and one that is unquestionably floundering; the suggested approach should not be directed at the pure science of other fields. Although the cautious review of any new scientific information is praiseworthy, there will be diminishing returns if, as a set policy, a witch-hunt/knee-jerk approach is adopted by our scientific community peers. We cannot afford to thwart original and novel efforts to bring about better science. Thus, our initial thinking when reading a submitted or recently published article's title or abstract should not be: "What on earth were these researchers thinking?" or "Who doesn't know that this is an inappropriate way to measure such and such?"

With only the brief review of the lizard-exposure study design provided just above, the reader would do well to anticipate what follows in the next pages, perhaps with his/her anticipation extending to identifying the most troublesome study design element of all.

- Are reptile populations decreasing at military installations, where relatively high explosive residues are seemingly accumulating? It is very doubtful that we know of such a trend. Consider how uncommon is it that military installations census their reptiles. Consider, too, that if censusing should occur, almost assuredly the frequency of census events is too low to reveal a phenomenon of plummeting numbers. A third consideration—if populations were, in fact, documented as decreasing, published papers reporting chemical effects of laboratory-dosed lizards would undoubtedly mention the population declines in their introductions, but we don't find this. We have, then, a first challenge to the study work; we don't know if reptile populations at military installations are imperiled, and this suggests that the essential study interest, of establishing a NOAEL or a LOAEL for a given explosive, has not been authenticated. We can go further, for we have now triggered no less than seven additional challenges to existing reptile ecotoxicology studies intended to support ERA.

- A more basic question than the previous one: Are reptile populations declining *worldwide* such that we should harbor grave concerns for species inhabiting U.S. military ranges? The above-cited fence lizard articles would have us believe so, but the troubling accounts of reptile declines in the principal cited source supporting the claim in the fence lizard articles (Gibbons et al., 2000) tell another story. Whether the cause of reduced numbers is habitat loss, the introduction of invasive species, environmental pollution, global climate change, or something else still, with perhaps only two exceptions, the examples hail from locations that are continents away. Thus, the bleak figures pertain to the common blacksnake (*Pseudechis porphyriacus*) of Australia, the Milos viper (*Microvipera schweizeri*) of Greece, the asp viper (*Vipera aspis*) of the Swiss Jura mountains, the short-headed legless skink (*Acontias breviceps*) and Eastwood's long-tailed seps (*Tetradactylus eastwoodae*) of southeastern South Africa, and the gecko (*Nactus pelagicus*) of the islands of Guam and Tinian.

- Military installations that conduct high-energy munitions and explosives testing are not amenable to reptile population censusing altogether! Commonly, the downrange portions of the installations where explosives residues accumulate are dotted with unexploded rounds that constitute a considerable safety risk, so much so that only specially trained individuals venture there, and only for the purposes of emplacing targets or retrieving certain munitions fragments and related information. Opportunities to establish that reptile/lizard populations are declining in these environments are therefore nil, and this begs a question: Are the explosives-dosing lizard studies that generate NOAELs and LOAELs being conducted only *on the chance* that downrange populations are declining? Sadly, this would appear to be so. Further, in that case, any resulting pronounced HQ exceedances based on limited soil sampling in the downrange areas would be particularly difficult to interpret. Consider that if populations should in actuality be declining—something that realistically we won't ever be able to verify—those aligned with the study work would have us believe that they have handily demonstrated a site's explosives residues to be responsible for the reduced animal numbers. This is an awful lot to take on faith. On the flip side, should downrange populations be stable—information to which we again won't ever be privy—those aligned with the studies would have their work cut out for them in trying to explain any pronounced HQs that would (likely) be generated.

To this point we have only considered one HQ consequence, that where constructed values are well in excess of 1.0. If explosives-based reptile HQs for a given munitions range should be at or below 1.0, though, those aligned with the study work would have us believe that, for the downrange area of interest, populations are not impacted. It is time to temporarily halt this discussion to conduct a reality check. The impetus for crafting reptilian NOAELs and LOAELs is the supposition that lizard populations are decreasing at military installations. The toxicological bases of the fence lizard NOAELs and LOAELs, however, include differences in body weight, organ weight, food consumption, blood chemistries, and testosterone concentrations. At best, then, HQs greater than 1.0 will only suggest that fence lizards bear skewed organ and body weights, altered or erratic feeding patterns, and other measures contrary to the norm. Who said anything about mortalities? Thus, researchers assuming that lizard numbers are decreasing have nevertheless opted to assess, in some crude fashion, whether or not a brief list of somatic measures are different in downrange animals relative to animals in relatively pristine environments. Are skewed organ and body weights, food consumption patterns, blood chemistries, and testosterone concentrations unhealthful for a fence lizard? If we imagine that they are, how much skewing signifies an impact of concern? Readers of the author's previous book (Tannenbaum, 2013) will hopefully recall here a particularly relevant discussion on second-order toxicology.

- Even if we assume that reptile populations situated in the downrange portions of military installations are truly in decline, and that NOAEL-based HQ exceedances have *bona fide* meaning and utility, why would anyone want to know if explosives accumulations were the cause of the decline? The military is not about to reduce its usage of explosives for any reason, although ongoing research initiatives to arrive at safer/more environmentally gentle explosives (i.e., so-called insensitive explosives) is recognized. Even in that case where a demonstrated linkage of accumulated explosives in environmental media and reduced reptile populations would become evident, the military is not about to cease or adjust its munitions-testing mission, and certainly not in order to protect reptiles inhabiting its ranges.

- Assuming that explosives residues of interest have accumulated in testing range soils, for how long has this been the case? The answer should be decades. This would suggest that if the residues can act to decimate populations, the decimation has already occurred. What purpose then, is served by deriving TRVs—for endpoints other than survival, please don't forget—at this late date?
- There is a last consideration regarding supposed worldwide reptile population declines providing the impetus for fence lizard studies aimed at generating NOAELs, etc. Those earlier-mentioned studies gave, as their justification for taking up the work, the phenomenon of explosive residues appearing in soil and other environmental matrices. Perhaps accumulations of explosives residues do account for some reptile die-offs in some discrete locations, but through simple deduction, one thing we confidently know is that explosives could not be accounting for any *worldwide* declines. Explosives residues in soil are highly uncommon. They occur only at the downrange locations previously described, and perhaps too at factories that manufacture fireworks if these should have significant spillages outdoors. If worldwide reptile declines should be a reality, explosives cannot be playing a role in this, even a minor one, for few and far between are the places where the residues occur. And so, the stated justification for researching explosives toxicity in lizards (seemingly as representative reptiles) just isn't there.
- There are just so many military installations that conduct high-energy munitions and explosives testing. While, admittedly, hundreds of acres at each installation belonging to this limited category may bear an explosives residue footprint, how shall we suppose the cumulative so-affected acreage—assuming that it is otherwise fence lizard-compliant—compares with the total acreage of suitable fence lizard habitat in the United States? Recalling that the installations conducting the munitions testing work are separated by vast geographic expanses (i.e., being distanced often enough by one or more U.S. states at a time), have we truly a case of concern? That is, do soil-bound explosives constitute a potential reptile threat such that fence lizard-dosing studies are necessitated?
- How many fences are there in the downrange portions of high-energy munitions and explosives-testing military installations? There probably aren't any. Any occasional fences that might have existed have long-since disappeared, having been demolished by the direct strikes of bombs and other projectiles, a cratered topography that cannot support upright structures, or the repeated forceful pressure waves associated with munitions-testing campaigns. For the recommended deliberate critiquing of wishful toxicology studies in support of ERA for the purposes of exposing fatal flaws and the like, we have now levied the most direct strike of all. Fence lizards have their appellation because they spend much of their time perched on fences, as well as tree trunks, and other

elevations, and with good purpose (Langkilde, 2009). These heights are sought out as vantage points from which to track prey and to scan broader portions of their immediate environs for potential predators. It is worth noting that behavioral ecologists have endeavored to document and quantify the lizards' dominant time allocation to available perching sites (relative to their time spent on the ground surface). So it is that while the previously cited papers have championed the fence lizard as a viable laboratory test species in support of research interests for health effects development in explosives-exposed reptiles, readers have been misled. Two disparate components of ecotoxicological study have been confused, namely the hardiness and adaptability of the species for indoor research, and the lizard's demonstrated suitability whereby its ecological role can be replicated to a respectable degree. The papers cite fence lizards as easily bred, able to thrive in a laboratory environment, having fast maturation and high fecundity, and being fairly small, such that large numbers can be housed in a relatively small working space. While these are all valid claims, it is highly unlikely that fence lizards (assuming they exist on bombing grounds altogether) would have daily opportunities to lap up explosives from small pooled collections (that form pursuant to rain events, we imagine?), that occur in recesses of logs or in depressions of other above-ground fence-like projections, that in any way parallel the forced-exposure slurry pipetting system described earlier. This is simply the wrong species to be testing with. While due attention has been given to ensuring a study design has incorporated essential toxicological tenets, it is the ecology that has been skipped over.

Through a review of two experimental study designs, the foregoing discussion has handily taken well-intentioned ERA-oriented research to task. There is no need to continue with a comprehensive review of a great many other studies, where, in similar fashion, weaknesses or all-out failings to inform ERA can be uncovered. The reader should be primed, though, to apply a simple two-tier "study utility assessment screen" for whenever the need might arise. En route to its implementation, one must first recall that ERA work need not mainly, if at all, involve the derivation of TRVs and the like. While toxicology is often called on for ERA support, ecological as opposed to toxicological considerations need to garner the primacy of investigations if ERA's lot is to be improved. Where published or planned studies are suggested for assisting with case-specific or generic ERA matters, the screen's first tier should ask if the study's premise is sound, specifically inquiring: "Does the study's ecological [risk] concern truly exist?" The second-tier screen asks if the science holds up. Here we are looking to see that the unique ecology of the study interest is being monitored or tested. A brief checklist would evaluate the degree of conformity of the experimental exposure or observational scheme to that occurring in a real-world setting. The reader should appreciate that, for the two study designs now thoroughly reviewed, both perform miserably when put up to the tiered screening. They are stuck in the "What would happen if" mode, and are not attentive to an ecological concern context that we know to exist. It is somewhat unfair to suggest that the PFOS egg-injection studies were without worth, but, in truth, it does not appear that birds are routinely being culled from the wild for liver PFOS monitoring, or that the derived PFOS LOAELs are still being applied. For our needs, it was never established that bird populations are declining due to PFOS in the environment, or that bird populations are declining altogether. As for the fence lizard studies, a valid need to learn of the species' response to the oral administration of explosives was never articulated. We don't even know if lizards are present on high-order munitions-testing ranges, and we're not likely to ever find that out.

Clearly there is a need to grow the science in support of ecological assessments for contaminated sites, and this book's purpose is to facilitate that endeavor. The three sections that follow are of grouped one-of-a kind studies. These recognize not only that efforts are needed in different venues—the field, the lab, the office work space—but that the proclivities of ERA enthusiasts can lead them to locations at least that varied. Within each section, an effort was made to methodically arrange the studies. The reader need not trouble himself/herself so much with this, for the study titles make evidently clear what each intends to pursue. Aside from the by-venue categorization, terrestrial studies are placed before aquatic ones, and only for organizational purposes (i.e., by no means a commentary that terrestrial gains for the ERA field are needed more so than aquatic ones; perish the thought). Additionally, and as could best be done, the studies are paced. Earlier studies within each section (particularly for the field studies) tend to get at substantiating or validating ecological risk concerns that often receive attention, while latter studies deal with unearthing mistakes and mismeasures, and with the active acquisition of new information. (Whatever the study kind, all necessarily begin with an active verb; we cannot sit back and expect good science to just happen by itself.) The earlier studies, then, recall what you have read about in the way of munitions-mediated exposures that are supposedly devastating lizard populations. The studies calling for accessing new and hopefully utilitarian data recall the PFOS-in-bird egg studies you read about. Those, you are reminded, could do no better than trigger guilt-by-association thoughts, and their one-time crafted LOAELs are likely just gathering dust today.

The cautious reader will have already secured the understanding that a bettered ERA will not follow from the continued use of the syringe, an intubation apparatus, or any other gadget designed to artificially deliver a chemical to a test animal. There have been far too many animal dosing studies of various kinds, and this book is certainly not going to add to the problems these studies commonly engender. (That said, for the purpose of securing a potentially great gain relative to our understanding of the chemical exposures of animals, there is one study of the compendium that involves a fair amount of dosing. Please allow me that singular exception, and please be on the lookout for that particular study. You are forewarned.) Importantly, and as was briefly mentioned before, dosing studies speak to toxicology, but it is the ecology we need to learn more of. Kindly do us all proud and tackle some of the studies you will read about. There's so much work to do.

References

Brunström, B., Axelsson, J., Mattsson, A, Halldin, K. 2009. Effects of estrogens on sex differentiation in Japanese quail and chicken. *General and Comparative Endocrinology*, 1; 163(1–2): 97–103. doi: 10.1016/j.ygcen.2009.01.006. Epub January 23, 2009.

Gibbons, J. W., Scott, D. E., Ryan, T. H., Buhlmann, K. A., Tuberville, T. D., Metts, B. S., Greene, J. L., Mills, T., Leiden, Y., Poppy, S., Winne, C. T. 2000. The global decline of reptiles, déjà vu amphibians. *Bio-Science*, 50: 653–666.

Houde, M., Martin, J. W., Letcher, R. J., Solomon, K. R., Muir, D. C. 2006. Biological monitoring of polyfluoroalkyl substances: A review. *Environmental Science and Technology*, 40: 3463–3473.

Langkilde, T. 2009. Holding ground in the face of invasion: Native fence lizards (*Sceleporus undulatus*) do not alter their habitat use in response to introduced fire ants (*Solenopsus invicta*). *Canadian Journal of Zoology*, 87: 626–634.

McFarland, C. R., Quinn, M. J., Bazar, M. A., Remick, A. K., Talent, L. G., Johnson, M. S. 2008. Toxicity of oral exposure to 2,4,6-trinitrotoluene in the western fence lizard (*Sceleporus occidentalis*). *Environmental Toxicology and Chemistry*, 27: 1102–1111.

McFarland, C. R., Quinn, M. J., Bazar, M. A., Talent, L. G., Johnson, M. S. 2009. Toxic effects of oral hexahydro-1,3,5-trinitro-1,3,5-triazine in the western fence lizard (*Sceleporus occidentalis*). *Environmental Toxicology and Chemistry*, 28: 1043–1050.

Molina, E. D., Balander, R., Fitzgerald, S. D., Giesy, J. P., Kannan, K., Mitchell, R., Bursia, S. 2006. Effects of air cell injection of perfluorooctane sulfonate before incubation on development of the white leghorn chicken (*Gallus domesticus*) embryo. *Environmental Toxicology and Chemistry*, 25: 227–232.

Ottinger M. A., Quinn M. J., Lavoie E., Abdelnabi M. A., Thompson N., Hazelton J. L., Wu J., Beavers J., Jaber M. 2005. Consequences of endocrine disrupting chemicals on reproductive endocrine function in birds: Establishing reliable end points of exposure. *Domestic Animal Endocrinology*, 29: 411–419.

Sulik, K. K., Johnston, M. C., Webb, M. A. 1981. Fetal alcohol syndrome: Embryogenesis in a mouse model. *Science*, 214(4523): 936–938.

Suski, J. G., Salice, C., Houpt, J. T., Bazar, M. A., Talent, G. 2008. Dose-related effects following oral exposure of 2,4-dinitrotoluene on the western fence lizard (*Sceleporus occidentalis*). *Environmental Toxicology and Chemistry*, 27: 352–359.

Tannenbaum, L. V. 2013. *Alternative Ecological Risk Assessment: An Innovative Approach to Understanding Ecological Assessments for Contaminated Sites.* Wiley Blackwell, Sussex.

Tannenbaum, L. V. 2014. A Fervent Plea for Second-Order Toxicology. *Environmental Toxicology and Chemistry*, 33: 479–480.

Winter, V., Elliott, J. E., Letcher, R. J. 2013. Validation of an egg-injection method for embryotoxicity studies in a small, model songbird, the zebra finch (*Taeniopygia guttata*). *Chemosphere*, 90: 125–131.

Section I

Field Studies

Study #1

Document Chemical-Induced Ecological Impacts (of Import) at Contaminated Sites

Premise

The terminus of countless ERAs for contaminated sites has been the reporting of NOAEL- and LOAEL-based HQs for multiple species that (often greatly) exceed unity. For avian and mammalian terrestrial receptors, the simple biological meaning of the exceedances is that the site receptors are, on a daily basis, consuming chemicals with toxic properties that are great multiples of safe- or effect-level doses. By all reason, whether the associated toxicological endpoint is reproduction or something else, animals at sites that bear these HQ features should display overt (visible) signs of compromised health and, more realistically, should be incapable of surviving. Reinforcing these anticipations is the reality that sites have been contaminated for three or more decades, and hence have all the while been imposing a continuous and unrelenting chemical stressor environment to site receptors. To apply the vernacular of many an ERA practitioner, contaminated sites have had all the opportunities they could ever need to have elicited *death, doom, and destruction*.

The suggested study can entertain myriad design possibilities. Common to all of them, though, is a deliberate effort to identify compromised animal health or more severe forms of ecological impact, the latter of which avid investigators are free to define and characterize.

Study Guidelines

1. Although it should be obvious, the study necessarily entails fieldwork. Potentially, biological or inanimate media samples gleaned from the field could submit to analysis in the laboratory (e.g., anatomical measurements), but study enthusiasts should duly note that significant

differences observed between or among them, collected at one or more contaminated sites and one or more non-contaminated comparison areas, will not illustrate impact. Importantly, too, modeling should play no role in the conduct of this study. The study is rooted in understanding the current ecological health state of a site, and projections of ecological interferences and the like are not being sought.

2. Demonstrations of impact do not extend to contaminant-level measurements in plant and animal tissues.

3. The study is an ecological one, and the collection of chemical concentration data in site environmental media, other than for the purpose of learning where concentrations may be greatest at a site, is to be discouraged. Those up to the challenge should plainly understand that the work to be completed is not a risk assessment. To the extent that one taking on this study would intend to integrate site chemistry into his/her analysis, ample contamination footprint information very likely exists already.

4. Interested parties must develop and consistently apply a defensible understanding of the phenomenon of "ecological impact of import." At a basal level, this study is seeking documentation that ecological function has been compromised, information that admittedly is not easy to gather. Going further, this study is seeking documentation that, without intervention, the site ecology is fated to become seriously or completely undone. This, then, would be documentation that will make it evident in retrospect, that listing the site and it having submit to some standardized assessment scheme was purposeful. Identifying that certain nutrient cycling at a site is reduced by 11% due to contaminant influences would only capture what the subject study is after, if the nutrient cycling deficiency was working its way to ecological collapse. Interested researchers are reminded that the longer a site has been contaminated by the time an observation such as this one has been made, the less likely it is that the ecological condition is degrading further. As should be clear, just because we have the capability to measure differences in ecological function (at a chemically contaminated site) does not necessarily mean that an ecosystem or the organisms that comprise it are aware of the differences. A discovered area of stained soil is not an ecological impact of import unless ecological site function is compromised at that location or elsewhere because of the presence of that location. Skewed enzyme or hormone levels are not impacts of import if organisms with these shifts are occupying the site and acting as they should. As one suggestion, a top-down analysis could prove to be quite useful to those with piqued interest in the topical area. In this approach, one is asking about what organisms or biological processes are lacking at the contaminated site, and how such is evident. One might ask: "If I didn't know I was at a chemically contaminated site, would I suspect that something, ecological function-wise, had gone awry here?"

5. The reality that baseline (historical) site population size information almost assuredly will not exist for a site does not have to mean that clever and inventive means cannot be developed and applied to produce workable findings. (It is recognized that lacking historical records of both the number of species utilizing a site, and onsite species population sizes, may deter interested would-be investigators.) Military installations quite commonly support an environmental office, and species lists, to include the documentation of special-status species, are often available for at least certain portions of these. Those electing to conduct their work at military installations should recognize two distinct features of sites on these properties. First, they constitute a noteworthy minority of the universe of contaminated sites in the United States. Findings brought forward, therefore, may not be descriptive or reflective of the lion's share of contaminated sites. Additionally, gaining access to locations of interest may either not be feasible at all, or may be restricted in ways that preclude getting at the valuable and specific information being sought.

6. The reality that portions of contaminated sites may have undergone intrusive remediation efforts (e.g., soil excavation) need not necessarily put such sites beyond the realm of consideration for use in a study effort. Intrigued investigators should consider the consequences, both positive and negative, of electing to work only with sites that have yet to advance to the remedial decision or remedial action phase.

7. Given the challenge of the study and the complexities that may be involved, the effort cannot be limited to a single contaminated site. While the scope of the study is not dictated here, it goes without saying that study investigators must defend the number of sites they considered in their evaluation. The study is therefore ripe for interested investigators networking to share ideas, resources, and study design approaches, with an intended aim at pooling their findings (and thereby building up the tally of sites investigated). Perhaps more than any of the other studies described in this book, this one lends itself to "competing" interests taking to the challenge of identifying sites with ecological impact, that as stated above, is "of import." This is not unlike a venture in the 1990s whereby multiple international modeling research groups took to the task of estimating average ^{137}Cs concentrations in a host of biological matrices and compartments in the aftermath of the Chernobyl accident. Although that venture involved modeling and the study described here has nothing to do with such, the subject study could nevertheless vie for also being termed a multiple participating party "test," one that enlists participants to generate better findings than those of "competitors," and, more importantly, findings that come closest to the real-world condition. Since the participation of multiple parties is needed to ensure that a sufficiency of sites are reviewed, enthused ecologists would do well to

explore what would be involved in arranging for a professional society or consortium to oversee and facilitate the coordinated contest of sorts described here. A formal tasking to contestants would articulate more detailed guidelines than are provided here, in addition to milestones to be achieved, and a terminal date for submitting findings to a centralized review panel.

8. Logic dictates that the multiple sites comprising an evaluation should share some common elements, and seemingly the more shared elements across sites, the more robust will be the findings. Universal "themes" could take the form of a focus placed on a singular animal group (e.g., birds), or on one specific ecotype (e.g., savanna).

9. Logic dictates that multiple tests, measurements, and metrics be used to buttress a study's findings rather than being reliant on a singular data type.

10. Investigators will need to be ever-cognizant that the status quo in ERA is the generation of HQs for certain receptors serving as surrogates for members of larger groups (e.g., a feeding guild). Present-day ERA determinations go no farther than opining about the health status of receptors; they do not discuss ecological functions, or potential or actual changes in ecological site dynamics. ERAs, then, only intend to generate conjectural surrogate organism-specific health statements. Examples would be stating that shrews will be/will not be reproductively impaired, or that earthworm survivability will be/will not be reduced. Not only do ERAs intend to make these statements, but ERAs are not capable of doing anything more refined that this. The suggested study's quest for the demonstration and documentation of ecological impact encourages the investigator to necessarily identify earmarks of an ecological upset condition at a site. Realistically, these earmarks extend to more than just noted population (size) shifts. In this context, investigators should be wary that site size might not allow for the illustration of decided population increases or decreases relative to the situation at comparable nearby areas, or relative to what may be reported in the open literature. Investigators should also be sensitive to the novelty of the charge given them in taking on this study. They will be endeavoring to identify a potentially unacceptable health state for a contaminated site, cognizant of the fact that present-day practitioners never dabble at a level of sophistication beyond the simplistic HQ construct. Investigators are additionally advised to be attentive to confounding factors that stand to compromise what they may believe to be indisputable proof of earmarks of chemical stressor-posed ecological impacts. Physical site alteration that may have occurred in a site's past is perhaps the best example of this, and possibly the one that will most frequently be encountered. Investigators should be wary, then, when declaring that

they have identified chemical-induced ecological impacts, because habitat disturbance might (more correctly) be the cause.

11. Intrigued investigators must be willing to address the matter of site size dynamics, and manage their efforts accordingly. Most contaminated sites that are the substance of remedial programs of the Superfund genre are, relatively speaking, small. While sites that extend to several hundred acres may appeal to would-be investigators, efforts taken to focus only on sites of this kind should acknowledge their lesser utility in terms of the study's goal of documenting instances of ecological impact at contaminated sites. (The lesser utility mentioned here is born of the concomitant expanded uncertainties of larger sites, and too, larger sites not reflecting the mainstay case of the Superfund-type site.) This is not to say that the documentation of ecological impacts at relatively large sites is to be avoided in the overall flushing out of the subject study interest. Ultimately, information is being sought for all types of sites as they may vary in size, former and present usage, nature of the chemical footprint, affected media, and more.

12. Vis-à-vis the study topic, evaluations of aquatic sites are just as necessary as those of terrestrial sites. In line with several of the preceding guidelines, interested parties should understand that what may be acceptable fare in present-day aquatic assessments will not be workable here. Askew indices born of comparing on-site measurements to tabular values for aquatic organism protection (e.g., sets of screening benchmarks for sediment-dwelling species) do not indicate ecological impact. The same is true for standard measures that purport to depict excessive pollution (e.g., chemical oxygen demand assessed in order to gauge amounts of organic compounds in lakes and rivers). It should be clear that traditional toxicity testing should play no role in demonstrating ecological impact; this study is not seeking to know how commercially available, laboratory reared organisms react to site water or site sediment.

Study Outcomes and Applications Thereof

Concerted efforts invested in this study may reveal that no evidence of ecological impact could be found, or that the frequency with which demonstrated impact is encountered across a goodly representation of sites is extremely low. Such outcomes could provide the substance to back initiatives calling for ERA reform. Where identified cases of impact are few and far between, the point to champion is that, since we are not seeing serious (or any) impingements in site ecology, there is no value added to continuing to conduct what we term baseline ERAs and the like. Concerted efforts could reveal that the required level of field study sophistication to allow for the identification of impacts is simply too resource-intensive to be practical. Such a finding would support another monumental gain for the field. It would underscore that, with the minimalist efforts routinely applied in ERAs, we know *a priori* that ecological impacts of worth (import) won't be discovered at sites sitting in the queue. In that case, relinquishing the bias and insistence that assessments must nevertheless proceed should pave the way for acknowledging the purposeless and valueless nature of assessments altogether, or at least until such time that utilitarian assessment schemes are developed. Should a secured conclusion be that sites are too small to enable the identification of ecological impacts of import (a reasonable expectation), such could be channeled in a beneficial manner as well. ERA reform here would amount to adjusting regulatory programs to dispense with evaluations of any sort for this dominant site category, with the overarching gain being the ability to allocate precious and limited resources towards environmental issues where energies invested could lead to environmental enhancements in other program areas. The common denominator to the above-described anticipated findings is the discovery that it is pointless to assess sites when, for one reason or another, we are aware that we haven't the ability to identify ecological impacts.

Study #2

Explore the Need for, Veracity of, and Utility of Toxicological Benchmarks for Non-Earthworm Soil Organisms

Premise

The suite of tasks that fall under the catch-all of "ecological risk assessment" for terrestrial concerns is finite, and by no means extensive. Ordinarily, soil contaminant concentrations are reviewed, a list of receptors-of-concern (for all intents and purposes, of birds and mammals only) is developed, and a HQ exercise (often scaled-down in preliminary go-rounds) is conducted. Less routine, but occurring frequently enough, is the conduct of toxicity tests that, within a laboratory-controlled ambient setting, pit commercially available organisms (e.g., earthworms, lettuce seeds) with site soils. There are also tasks that never occur in support of terrestrial ERAs, a matter that gives rise to this study. Site soils and the litter layer are never examined for the presence of enchytraeid worms (potworms; these can number as many as 250,000 per square meter of ground surface), or springtails (hexapods [no longer considered insects] of the phylogenetic subclass *Collembola*, measuring no more than 6 mm in length). Mites and woodlice (pillbugs) as soil invertebrates other than earthworms are also not investigated at the soil surface, and the same is true for the nearly endless list of soil microbes that exist in the soil surface microenviroment. This leads to a great curiosity, for while these species are never evaluated in the field (for presence/absence, overall health, counts), they are, among others, common test organisms for the development of soil screening levels (SSLs) and for ongoing research to identify toxic effect thresholds, where both of these efforts are intended to support ERA. The curiosity is exacerbated through a review of the life histories of some of the SSL-development tests species. *Perionyx excavatus*, a composting worm with the common name "blues," is known for feeding on dung. Further, it is described in a key toxicological benchmark guide as "the

Indian subcontinent equivalent of *Eisenia fetida*," preferring compost heaps and other accumulations of organic material. *Pheretima posthuma* is indigenous to east and southeast Asia, and notably can grow to a length of 30 cm (12 inches), quite unlike many Lumbricids found in U.S. soils. No life history is available for another species, *Octochaetus pattoni*, although it is known that other species within its genus occur in New Zealand. The ecological roles served by potworms, springtails, mites, pillbugs, and microbial heterotrophs are extensive, seemingly vital, and well-recognized, with organic matter decomposition and soil aeration being primary ones. All that said, we should find ourselves beset with numerous, hard-hitting questions:

- Why develop soil and litter toxicological benchmarks if soil and litter invertebrates and microbial heterotrophs at contaminated sites are never surveyed or health-assessed?
- Why develop toxicological benchmarks for soil and litter invertebrates and microbial heterotrophs if, pursuant to predicting soil or litter benchmark exceedances, field verification of supposed impacts to the organisms involved does not proceed?
- Why develop the toxicological benchmarks through deliberate laboratory testing or extensive literature review if their basis reflects the abilities of species far-removed—both geographically and ecologically (the latter, for example, referring to dung not typifying the land surface at contaminated sites)—from the true site condition?
- Of what utility are the benchmarks if they are not being used/applied? (Ecological risk assessors should consider if they have ever come across assessments that either [a] commenced with screening on-site potworms and the like in order to construct COPC lists, or [b] concluded that these organisms bear unacceptable risks.)
- Of what utility are the benchmarks since we do not remediate sites to afford greater protection to soil and litter invertebrates and microbial heterotrophs (the "two groups" mentioned in the next paragraph and onward) that we may suspect are at risk? (The reader is reminded that exceeded soil invertebrate SSLs do not speak to the welfare of other/ higher trophic-level species within the same ecosystem.)

This study is intended to seed an open exploration into an ERA topical area that does not appear to command a great deal of attention. (The reader is reminded that screening for potential soil invertebrate effects, even to standard Lubricid earthworms, does not proceed as an early ERA step.) Through all of its undertakings, the study is directed at securing justification for evaluating the well-being of the two groups of organisms.

In brief, the necessarily multifaceted tasks are directed at understanding (a) the need (if there should be one) for assessing the groups, and (b) the value (if any) of observed benchmark (e.g., SSL) exceedances.

Study Guidelines—Empirical Work in the Field

1. Collect comparable soil and litter samples from contaminated sites and nearby habitat-matching non-contaminated (reference) locations, and assess populations of non-earthworm invertebrates (i.e., potworms, springtails, mites, pillbugs) and microbial heterotrophs at both. Ensure that the contaminated sites sampled embrace an array of chemical stressors, to include, at a minimum, metals and organic species. Identify associations of non-earthworm invertebrates and microbial heterotrophs, and contaminants. Endeavor to sample and identify associations at multiple times of the year.

2. For the contaminated sites, also tabularize the population figures and associated soil chemicals and their concentrations (range and arithmetic mean). To the extent that soil chemical concentration data exists for the staked locations of previous RI investigations, necessarily conduct the population assessments at these (thereby facilitating project cost savings through not having to run analytics anew). Add table columns listing the invertebrate and microbial heterotroph species identified at the reference locations. Document how long contaminated sites have borne their chemical release constituents.

3. Where contaminated areas reveal considerably lesser invertebrate and microbial heterotroph populations than do matched reference locations, document any aspect of the immediate site ecology that appears to be stressed or compromised at the gross observation level (e.g., dwarfed populations of small mammals, stunted or diseased vegetation). Determine if concomitant compromised soil microhabitat ecosystem function (e.g., organic matter decomposition rate) occurs at the contaminated areas.

4. If site contamination is found to appreciably reduce populations of non-earthworm invertebrates or microbial heterotrophs in soil or litter (guideline 2), arrange to create this situation anew in the field. Identify comparable non-contaminated (reference) locations and, pursuant to censusing their baseline invertebrate and microbial heterotroph populations, apply commercially available chemicals to them, to arrive at concentrations akin to those at the contaminated sites. Assess invertebrate and microbe populations at the newly created sites and the long-time contaminated sites 2–4 times per year for 1–2 years.

5. Through novel means (perhaps through accessing an experimental forest, or use of irradiation), arrange to extirpate soil and litter non-earthworm invertebrates and microbial heterotrophs at a series of non-contaminated locations, preferably not impinging on earthworms in the process. (If earthworms as non-target organisms should suffer die-offs, attempt to repopulate the locations by transferring earthworms from non-treated areas, being careful to shield them from ultraviolet light by placing them below the soil surface or beneath litter cover.) Artificially manipulated locations should be sufficiently large to allow for the observation of any changes in ecosystem function, generally but not limited to food-chain dynamics (e.g., the presence of various feeding-guild members). Monitor successfully transformed locations over at least two years, documenting all observable and quantifiable changes to birds and mammals (and potentially, too, reptiles and amphibians). For small rodents (and potentially other mammals as the locations will permit) and birds, tasks can include recording overall health and appearance, censusing populations, measuring animal activity levels, and analyzing stomach contents.

Study Guidelines—Laboratory Study

1. Assess the suitability of *Perionyx excavatus* as a surrogate for *Eisenia fetida* in standard testing for the development of toxicological benchmarks for soil invertebrates. In particular, investigate the relative thresholds-for-effect for the two species. Establish why *Perionyx excavatus*, a species that is not indigenous to the United States, would be used in establishing soil benchmarks or in SSL development, given that *Eisenia fetida* remains so commercially available.
2. For several metals and organic compounds (to be tested individually), replicate standard earthworm toxicity testing with *Eisenia sp.* (or other Lumbricid) and *Pheretima posthuma*. Investigate relative thresholds-for-effect to assess the surrogacy capability of *Pheretima sp.*
3. Develop one or more toxicity tests that are fully predictive for the field, with the tests targeting organism survival and/or a valued ecosystem component that stands to be impacted from an organism's reduced health or numbers.

Study Outcomes and Applications Thereof

With all of its elements considered, this study should primarily allow serious ERA practitioners to recognize misdirected ("loose-cannon") efforts that populate the ERA landscape often enough. At the very least, the discriminatory abilities of ERA practitioners stand to be augmented in the following way. It is easy enough to come to readily accept novel ecotoxicological information as being only to the benefit of ERA, and particularly where the research delves into a more obscure and specialized area. As just one example, attention to soil and litter microbial heterotrophs does not, by any means, describe a key focus of ERA. Yet, when researchers go on to produce a benchmark set or a guidance document for the evaluation of this terrestrial ecosystem component, it can easily come to be accepted—and regularly implemented—at face value. Where, though, is the demonstration that the developed screening benchmarks are workable? How important could screening these families of organisms be if it has been ignored for the first several decades since ERA's inception? Do benchmark exceedances actually translate into compromised ecological function in the field, and particularly for birds and mammals (the only two vertebrate groups routinely studied)? Importantly, the infatuation that breeds over the prospect of having a new ERA area to latch onto can extend to negligence with regard to validation of disseminated information. Producing a toxicological benchmark compendium is likely premature if it reflects no more than having combed the open literature for documentation of chemical effects. Given that sites that submit to ERA have been contaminated for decades, a compendium isn't truly helpful where the testing methodology is far from standardized, and where test durations might be as brief as 2.5 hours or 7 days for microbial heterotrophs and nematodes, respectively. The study, it is hoped, will prompt ERA practitioners to question the relevance and applicability of reported testing approaches (e.g., culturing nematodes in microtiter plates) and developed benchmarks, to the actual terrestrial site condition. Pursuant to involvement with the subject study, the astute ERA practitioner will come to better appreciate the disclaimers of benchmark compendia. Of note, toxicity benchmarks for soil invertebrates bear the same disclaimer published with plant toxicity benchmarks, namely that a benchmark is a poor measure of risk where soils are observed to support high numbers of the organisms being screened.

In an overall sense, study outcomes may foster utilitarian discussion on the benefits of streamlining ERAs. ERAs have no need to gather and process information where certain tasks can be relegated to the "no value added" category. Additionally, where tasks can be removed, streamlined ERAs can pave the way for precious environmental science resources being applied to issues deserving of greater attention.

In the main, interested parties are asked to research two central notions. Because birds and mammals in terrestrial systems could be indirectly harmed, there is a need to know if there are observable differences between contaminated sites and habitat-matched reference locations with regard to the quality of soil and litter organisms. Where site contaminants have impacted soil and litter organisms (assuming such can be demonstrated), there is a need to learn of the degree to which the imposed changes have disrupted a contaminated site's greater ecology.

Study #3

Demonstrate the Erroneous Nature of Avian and Mammalian TRVs through Hands-On Animal Study

Premise

Not uncommonly, computed HQs for avians and mammals at contaminated sites exceed unity (a value of 1.0) by a considerable degree. Not uncommonly, too, these large HQs lead directly to site remediation thoughts and discussions focused on re-tooling sites such that they can sustain their ecological species in a chemical stressor-free or stressor-reduced environment. Repeatedly overlooked in the site-management context is that the computed HQs of seeming concern are outright impossibilities; no avian or mammal can remain alive for any period of time if it bears, for example, a HQ of 79. Still higher HQs of say 112 or 304 (that the author, and no doubt the reader as well, has come across in professional work), in addition to being impossible, bear on the whimsical and the absurd. We understand that chemical concentrations in soil or in dietary items directly fuel the HQ magnitude, and that, everything else being equal, HQs developed from site areas with relatively higher contaminant concentrations will be the ones to give rise to a site's relatively higher HQs. Is there, though, any onsite avian or mammal with a sustained HQ of 79? There could not be; by definition, an animal with a HQ of 79 consumes, each day, a chemical with toxic properties at 79 times the safe dose! (The author, for obvious reasons, will refrain from discussing excessive HQ magnitudes, and particularly those like 79s [again] that are LOAEL-based, and where mortality is the associated toxicity endpoint.) For how long could this sort of dietary practice go on?

If a receptor consumes a chemical at that rate, it would have long been erased from the site. (The reader is reminded of Study #1 that encourages interested parties to document contaminant-driven ecological impacts.) But the modeled receptor as a guild surrogate must have been present at the site nevertheless, for why else then was such an animal evaluated altogether? ERAs are not in the habit of evaluating hypothetical ecological receptors, as if to ask: "What would the hazard be to species X if it occurred at this site?"

With the HQ construct so simplistic, tracing the source of "HQs of impossible magnitude" presents little challenge. For all intents and purposes, we deal with a finite list of avians and mammals, and we know of their diets, both in terms of what they are comprised of and in what quantities diet items are consumed. Granted, we could collect the dietary items (field mice for example, in support of a fox's assessment; earthworms and grub worms in support of an omnivorous rodent's assessment) to directly measure their contaminant levels, this translating into more accurate intakes than we arrive at through modeling the fate of soil concentrations. Still, this is not the source of the slippage in the crude hazard assessment scheme; the source of the greatly inflated HQs can only be the TRV. We can readily compile a list of factors that contribute to hazard mismeasure that follows from TRV usage. At a minimum, the list would include using animals that have never before been chemically exposed as test subjects, chemical delivery though the most unnatural of means and in a far more direct manner than a field animal would ever experience (e.g., via syringe, i.p. injection, contaminants laced into the [artificial] diet), and exposure periods of perhaps just four weeks. Let us observe, too, that cohort (i.e., one-generation) studies supply no meaningful information if site receptors have already produced tens of generations by the time an assessment is done. TRVs, then, only inform on how an experimental animal responds in its most unnatural milieu— in a cage, under fixed lighting and temperature conditions, force-fed, and with all of this (in the case of rodents, certainly) in genetically constrained subjects only. There are only two ways to explain (the presence of) the HQ-of-79 receptor; it is either more chemically resilient than we give it credit for, or the TRV supporting the HQ is faulty. When we observe animals with double- and triple-digit HQs at sites, we must not insist that the HQ is right and that our eyes are playing tricks on us (i.e., that we are not actually observing receptors in the wild, or clear signs of them); we must rather concede that the calculation is wrong. For the betterment of ERA, this study seeks to incriminate the TRV through assorted means.

Study Guidelines

1. Compile a list of contaminated terrestrial sites, as yet not remediated, for which excessive/impossible HQs for avians and mammals have been computed.

2. Working with toxicologists and animal physiologists, determine for a range of tissues (e.g., liver, spleen, bone, blood) and for the whole body approximate anticipated concentrations of a broad range of soil chemicals encountered by mammals and birds at Superfund-type sites. Necessarily determine the tissue and whole-body concentrations to manifest from exposure to areas bearing discrete soil concentration ranges. In working with metals, for example, consider soil concentrations up to 50 ppm, from 50 to 100 ppm, and from 100 to 500 ppm. See guideline 5.

3. Obligatorily, where one or more rodents (or, possibly, small insectivorous mammals; see note in next guideline) had excessive HQs, live-trap these animals at the site for direct observation. Record general health condition, and any overt signs of illness or disease (such as body sores, arching or other erratic behavior, altered pigmentation, and tremors).

4. Even in the absence of an observable compromised health condition, arrange to house some of the field-trapped rodents (see guideline 5) indoors, understanding that the sudden imposed areally restrictive condition and now-supplied laboratory diet could severely stress the animals and lead to compromised longevity. (Note that shrews, truly insectivores and not members or the Order *Rodentia*, will almost certainly not cage-adapt, much less survive a night in a confined live trap prior to conveyance to laboratory housing.) At nearby habitat-matched reference locations, field-trap approximately the same number of rodents of the species captured at the contaminated properties, with these to also be housed indoors. With allowances for the imposed and forced living arrangement, determine if contaminated-site rodents fare any less well than do reference-location rodents.

5. Euthanize some additional contaminant-site rodents and analyze the somatic tissues researched in guideline 2 for those site soil chemicals responsible for the excessive HQs. Compare the estimated tissue and whole-body burdens against the actual burdens of the site rodents.

6. Assuming field rodents can successfully adapt to cage housing, plan to dose reference-location rodents in a specific way. Revisit site-specific "excessive" HQs to determine the mg equivalents ingested daily (as, for example, the mgs of Chemical X for a white-footed mouse with a HQ of 79). Orally dose the rodents, matching the testing duration of the study that serves as the basis of the TRV. Endeavor to administer the chemical in small soil quantities, if possible, or in actual diet items (e.g., earthworms, seeds) in an effort to mimic the manner in which oral exposures occur for

the rodent (as occurs through preening and other incidental behaviors). Monitor health and activity levels for the duration of the study.

7. After securing animal care and use committee approval, trap contaminated-site birds and, at a minimum, draw off blood samples to be analyzed for the concentrations of the so-called "excessive" hazard drivers. As in guideline 5, compare estimated and actual tissue concentrations.

8. Following from guideline 3, trap white-tailed deer from contaminated sites (i.e., only in situations where these animals spend 50% or more of their time at these, this a mark of site fidelity), where these animals have recorded excessive-magnitude HQs. (Note that the exercise will be pointless unless the contaminated sites support a sufficiency of deer, perhaps 5 to 10 minimally, in addition to satisfying the site fidelity requirement. Recognize that, deer densities being what they are, and with the species' average home range being approximately 640 acres, sites any smaller than approximately 125 acres will not be suitable for this guideline exercise. Recognize, too, that assessing white-tailed deer, HQ-wise or in any other fashion, is purposeless for sites any smaller than 125 acres.)

9. Following from guideline 6, revisit the "excessive" HQs of white-tailed deer to arrive at mg equivalents ingested daily. After securing animal care and use committee approval, endeavor to orally dose the deer in outdoor/penned research facilities to see if animals succumb toxicologically from the treatments.

10. Endeavor to live-trap other mammals (e.g, rabbits) and birds at contaminated sites that recorded unwieldy HQs, provided that it was proper in the first place to have computed HQs. Thus, prior to any animal-capture work, verify that (a) the site is large enough to regularly house a sufficiency of animals of this type (perhaps a minimum of five), and (b) all animals spend 50% or more of their time contacting the site. At a minimum, record their general health condition as described in guideline 3. Endeavor to analyze tissues (minimally blood) for contaminant levels to support comparisons with estimated tissue concentrations. As with the described rodent and deer work, endeavor to collect from nearby habitat-matched reference locations species observed on-site. Dose these reference-location animals with the mg equivalents that drove the excessive computed HQs for the sites in question.

Study Outcomes and Applications Thereof

The information to potentially be brought forward through the assorted tasks is truly of a fascinating nature. An ERA has a standing obligation to understand why contaminated sites, all of which have been contaminated for more than three decades, can support the species they do. Study results will decide if body burdens are health consequential, and reveal, as well, whether or not assimilated quantities of chemicals are as estimated. To the extent that field-culled avians and mammals appear healthy, ERA practitioners will learn that HQs are often (always?) misleading indicators. Recalling that virtually all LOAELs are between five and ten times higher than NOAELs, and that NOAEL-based HQs of 20 (i.e., far less than "the poster-child HQ of 79") necessarily mean that reproductive impacts or mortality are certain, excessive HQs will have been shown to be erroneous. Moreover, the ERA practitioner will have full license to hold the TRV at fault for the lack of predictive or descriptive power of HQs. The study stands to demonstrate openly that the laboratory-dosing design of studies supporting TRVs simply does not align with contaminant exposures in the field. How is it that HQs commonly are unacceptable even for the site background (not that it should ever be standard practice to compute background HQs)? Where the study results are handily shown to be TRV-incriminating, they will constitute a wake-up call for regulators. First and foremost, this audience must concede that terrestrial sites are not stressing avians and mammals, the only regularly evaluated receptors in these settings, and that sites are not in need of cleanups. Second, regulators must move to put a permanent halt on conventional toxicity testing that spurs on TRV development. Third, regulators should encourage the development of improved screening and assessment tools, with an emphasis placed on demonstrating receptor wellness.

Study #4

Determine Required Minimum (Species-Representative) Bird Counts for ERAs at Prototypical Contaminated Sites

Premise

Guidelines on receptor-of-concern selection for ERAs at terrestrial sites are never well-stated. We seem to understand that selected species should be guild surrogates, reflecting the ecological niche of (multiple) other species that might populate the contaminated site of interest. Receptors-of-concern should also have (spatiotemporally) high degrees of site contact. This condition recognizes that animals with transient exposures are unlikely to be chemically challenged and, by the same token, allows (at least in theory) for noting when the potential for ecological impacts is more credible; the more time it spends at a site, the greater the chance that an animal will be at risk. It is said, too, that selected receptors should be keystone species, i.e., those that, because of their ecological role, could cause the local ecology to be further impeded if their numbers should be reduced. Finally, where possible, an effort should be made to include selected species that are of societal relevance.

The honest ERA practitioner will acknowledge that the general guidelines just reviewed are not all that helpful in the case of birds and, further, that ERAs for prototypical sites don't often (ever?) implement the guidelines for this animal group. We can demonstrate this with a consideration of the robin (*Turdus migratorius*), a frequently selected receptor-of-concern in terrestrial assessments, and one for which a basis for its selection as a surrogate for other birds is almost never supplied in ERAs. We would be hard-pressed to argue that the robin at a prototypical site is a keystone species. We first recognize that a robin, or a territorial pair of these birds, would actively exclude all other robins and other avian interlopers from occupying a space of a few acres (on average). Should chemical toxicity or some other human interference over a site's few acres be responsible for a

robin or a territorial pair being removed from the picture, no one can seriously argue that the area will cease to normally function in an ecological
sense. That is, there would be no anticipation that the site's earthworm
population would suddenly expand exponentially because one or two
vermivorous robins were displaced. There would also be no anticipation
that an imagined earthworm population explosion resulting from reduced
worm predation (born of the eradication of the bird pair) would translate into excessive soil aeration and/or hydration such that a long-standing
plant community would be eradicated. We should necessarily continue
with our taking the robin to task. Do ecological risk assessors recognize
that robins have an exceedingly high mortality rate, with up to 80% of
the young dying each year? As part of the natural ecology, tree squirrels,
chipmunks, raccoons, magpies, crows, ravens, and jays eat robin eggs and
nestlings. Thus, for a five-acre land parcel, the expectation of even so much
as a single robin pair might not be a fair one. Is the robin a good choice,
then, to serve in a surrogate role for other vermivorous birds, such as certain thrushes? Are we, for example, clear on how many different bird species, and how many individuals of each species, a nesting robin pair might
be representative of at a given five-acre parcel? Another consideration: the
American robin population is 320-million strong; there are as many robins in the United States as there are people! With such a presence, it's quite
unlikely that chemically contaminated sites, even as we consider them collectively, are making the slightest dent in the country's robin population.
Further, no one would ever notice that a given discrete contaminated site
is killing off, outright, the literal handful of robins it might contain. If, by
every measure, robins are going strong (i.e., with no indications of a population decline that would earn it International Union for Conservation of
Nature status), why should this species be the one chosen to represent others, and why should we be suspect that other birds of its guild would be
faring any less well? Additionally, we have no information to argue that
the robin is notably less or more chemically resilient than other birds in its
guild. Making the case that the robin is societally relevant is a hard sell. Its
reliable presence over a large portion of the United States is documented,
and its great numbers vanquish from anyone's imagination the possibility that it might one day disappear. While it is three times a state bird, its
image does not recall the American bald eagle or California condor of the
1970s, when these species were on the brink of extinction.

   While we find the decision-logic supporting choices of certain birds as
surrogates baffling, if not altogether absent, birds often present us with
another complication in a receptor-of-concern context. Because they
move about so much, it is hard to link a bird to a site in a meaningful

way, i.e., so as to be able to say that it "lives" there. Consider that nesting activity, though surely demonstrating a site presence, is a transitory phase. How many birds, then, of any one type must a site have/support for it to be rightfully evaluated, and how many other guild members need to populate a site to legitimize an avian component to an ERA? What if a site doesn't support other guild members? In such a situation, how can the commonly assessed receptor-of-concern be said to be a species surrogate? What if a closer look reveals that a site is regularly home to eight or ten species, but for all, or nearly all, of these there is but one representative? In such an instance would an ERA have an avian component to address altogether? ERA guidance won't supply the answers to these questions, and our starting point for getting at the critical numbers is that ERAs intend to evaluate populations as opposed to individuals. Beyond this, we know that a population minimally consists of two species representatives, and also that remedial actions do not proceed because there are anticipations (perhaps even sightings) of just two or a small handful of individuals at a contaminated area. This study invites interested parties to craft a standardized and defensible method for the rightful inclusion of birds in ERAs based on anticipated or actual numbers of birds occupying and utilizing chemically contaminated properties.

Study Guidelines

1. Intend to have the method or guide to emerge report on all regions and ecotypes across the United States, or across any other country that submits to study. As with a good many of the studies of this compendium, recognize that a meaningful and utilitarian product can only result from the aggregated findings of multiple interests.
2. Extensively review the literature to categorize birds by guild (i.e., feeding design). Refine the categorizations by state(s), sectors of states, and by habitat.
3a. For an extensive list of species, extensively review the literature to arrive at approximate bird and nest densities per acre or per hectare.
3b. Drawing on species-density information, home ranges, foraging ranges, and behaviors, cautiously endeavor to secure bird counts for 1, 5, 10, 20, and 30-acre parcels, for the situation where a bird is present on a parcel for a minimum of 45% of its time, discounting those months of the year when birds are overwintering at a distant location. Note that the 45% figure is not an arbitrary one; the figure reflects an understanding that the more time an animal (here, a bird) is away from an area (that could theoretically be a contaminated location), the less likely it is that it will

succumb to its chemical exposures. A bird that is present 45% of the time (as either the cumulative time of portions of days, or a percentage of [entire] days relative to a complete calendar year), is absent from the area of interest 55% of the time, and surely not enough to spark a legitimate ERA interest. (Researchers are invited to select a site presence figure lower than 45%, provided that they can defend such a minimal figure as allowing for birds to sustain noteworthy toxicological effects from site-specific chemical exposures.)

4a. For a given specific habitat within a portion of a state, randomly select five or ten 1, 5, 10, 20 and 30-acre parcels (that, realistically, could be chemically contaminated sites). At each, identify one or two species that trained field ecologists and wildlife biologists could agree make for an appropriate guild representative or surrogate. At each of the selected study parcels, identify all the guild members that correspond to the chosen surrogates.

4b. Alternatively, consider selecting the variable-size study parcels at or very near to sites that have had terrestrial ERAs done in recent years. See guideline 9.

5. Using multiple means (direct observation, camera work, other monitoring techniques), directly assess the total number of birds spending at least 45% of their time at the parcels, and the number of representatives of as many individual species as can be assessed. Recall that (a) mere bird sightings (e.g., fly-overs) are not ERA-relevant, and (b) typical contaminated sites, the subject of this book, are never so perniciously contaminated such that singular receptor exposure events are capable of sealing a bird's fate. Understand that to obtain the quality information sought, observations and counting should occur over the entire calendar year.

6. Categorizing birds by guild, tabularize the number of representatives of each species that are present a minimum of 45% of the time (or for some defensible, albeit lesser, figure; see guideline 3b).

7. Develop a user-friendly index or guide that in the main indicates when bird inclusion within an ERA is appropriate. Have the index segregate its reporting by geographic region, ecotype, and parcel size (guidelines 3b and 4). Where the guide indicates that assessment is appropriate, indicate if a given species would be likely to be serving as a surrogate, or if the assessment would be for that species alone. (Duly note that, where a given species could be a surrogate for one or more species in a given region, such may only be true for settings that are physically larger than those of typical contaminated sites.) Bird densities alone, for example, could actualize such a condition. Thus Species A, amply present at the four-acre Site X to warrant assessment (i.e., with hopefully substantially more than one male and one female regularly occupying the property) might be a guild surrogate for Species B, C, and D, but only where the area of contamination extends to 12 or 15 acres.

8. For terrestrial sites that are slated to submit to ERAs (across various regulatory programs), endeavor to have the index/guide indicate what bird(s) would be appropriate for inclusion, and which species, if any, could rightfully qualify as surrogates for others. (The latter recalls that due to site sizes and bird densities, certain species might not be represented by an intended surrogate despite being notably similar in terms of habits.) Such projections would, of course, likely only be based on (general) site location and site size information; it is understood that an enhanced fluency with study site particulars could yield different projections. To the extent that regulators and contractors would be receptive to the input, notify these parties of the gathered information that should be of assistance in the design of forthcoming ERAs.

9. Itemize all instances of specific birds of past ERAs having been inappropriate selections because they were not present in sufficient number. Itemize the instances in site-specific ERA applications where a bird designated as a surrogate was not truly so. Estimate the percentage of past ERAs in which one or more assessed birds were inappropriate selections. For past ERAs that included at least one bird, estimate the percentage that needn't have had such an inclusion.

Study Outcomes and Applications Thereof

This study is directed at bringing truthfulness in reporting to ERAs where birds may stand to be a concern. For the present, ERA seems to assign designations to ERA components without doing the legwork to show that the designations are accurate. Assigning a "surrogate" status to a bird is one case in point intended to streamline assessments; should the designee be found to be at risk, it may be safe to assume that other ecologically aligned birds at a site may also be. Assigning a "keystone species" status to a bird provides another example. We potentially stand to gain substantial knowledge if an ERA can make the case that a pivotal site species is being threatened. Before arriving at a level of sophistication that paints a bird as a surrogate or a keystone species, however, ERA has to firmly establish how many birds to potentially experience chemical impacts in site-specific instances are enough to validate avian assessments altogether. It is a fair statement that presently ERA practitioners operate with the thinking that including at least one bird in a terrestrial ERA is an absolute requirement, but there is no basis for such. The research leading to successful completion of this study will necessarily bring forward the information to know if a given bird has the force to represent others and, more simplistically, if the bird itself should be assessed. Where many researchers participate in the study, the findings should first put ERA practitioners in better touch with bird densities. Although ERA guidance and other compendia provide density information, we rarely if ever find ERA's applying the same in establishing the assessment's scope. The study stands to secure an important reality for the ecological risk assessor; while lots of individual birds may utilize a site, if there is scarcely more than a pair of birds for any one species, a case for avian assessment being essential is not made. Greater total bird counts cannot justify ERA attention to avian assessment if within-species populations are non-existent or so very minimal. In a broader sense, should minimum bird-count screening (for determining appropriate species for ERA inclusion) more often than not indicate that species do not qualify, such can lead to a greater good. Study findings can contribute to ERA reform, reminding practitioners that what were once thought to be mainstay assessment components are not really so.

Study #5

Field Document Spent Shot Pellet Ingestion by Grit-Ingesting Birds

Premise

That sizeable populations of grit-ingesting birds may be continually succumbing to poisoning, illness, and death from the accidental ingestion of spent (lead) shot pellets lying on the ground at shooting ranges has been an expressed concern for multiple decades. It is true that thousands and millions of visible spent pellets lying atop the soil may abound in certain areas, presenting a highly accessible potential risk to these species. It is also true that, on occasion, necropsies performed on dead birds found in pelleted regions have revealed the presence of a pellet or two in the gizzard, suggesting pellet ingestion as the likely cause of death. Despite these concerns, there are reasons to be doubtful that incidental pellet ingestion is wreaking such havoc. No bird population losses, let alone substantial population impacts attributable to the phenomenon, have been documented in the United States. Further, the more recent wave of ecologically relevant pellet-dosing studies, intended to better simulate these particle lead exposures that may actually occur, has shown birds to be highly tolerant to upset. Importantly, too, flawed assumptions have been identified for some well-intentioned models that endeavor to estimate pellet-ingestion event risk (by, in part, evaluating the relative frequencies of true-grit particles and appropriately sized shot pellets in soil samples). Given the above, it would seem prudent to establish the veracity of this (assumed) exposure pathway upon which so much ecological protection attention and research continues to be focused.

The suggested study, a field verification endeavor, aims to bring forward, in an unbiased fashion, the information to validate the phenomenon of birds often enough ingesting spent shot pellets. The study is necessarily a quantification effort for the behavioral phenomenon, with the specific goal of learning of the frequency with which grit-ingesting birds alight on the ground where significant stores of accessible pellets are present.

Study Guidelines

1. First and foremost, the field study can only proceed at locations where (ideally more than one) grit-ingesting species are well documented as being present. Suitable study locations are those where pellets are highly visible on the ground surface for most of the year, and where pellets are fully obscured from view only because of snow accumulation. Accumulated deciduous leaves on the ground surface only render a site non-suitable if the ground cannot be seen because the leaf layer is so thick, a condition that would make it unlikely for birds to bother with pecking at the ground in hopes of finding true grit particles anyway. Areas that are free of vegetation initially but then become overtaken by tall grasses and other plants (e.g., grains) that are periodically mown/cleared/harvested are not suitable study areas for a number of reasons. Aside from the often overgrown vegetation obscuring pellets and blocking birds from getting at the ground surface, the use of invasive equipment (e.g., plows) to clear areas will necessarily interfere with pellet deployment (e.g., through burial), and by unnaturally compacting soil. Pursuant to heavy rains, the latter can lead to prolonged periods of standing water, a situation inconsistent with grit uptake. The use of invasive equipment may also introduce a confounding factor for the study; should relatively low numbers of birds be observed at these areas, such might trace back to the noise and other habitat disturbances of the heavy equipment.

2. It should be clear that entire regions of the United States may not be able to supply qualifying sites. This observation can figure prominently into a consolidated understanding of the existence/scope of the potential problem area.

3. Suitable sites are those that have ample accumulations of pellets of the sizes 7.5, 8, and 9 (i.e., those pellets that approximate the size of true grit particles, generally ranging from 1 to 4 mm in diameter). The presence of shot particles larger than size 7.5, aside from a considerable representation of the pellet sizes that this study demands be present, is not problematic, for grit-ingesting birds do not entertain these.

4. Seemingly, the ideal study site is the fall zone of an inactive trap or skeet range, where, of course, the fall zone has not been swept of accumulated shot. The prospect of regular site accessibility and the presumed absence of human interferences (other than infrequent trespassers, perhaps) are the great advantages offered by this kind of site.

5. The study necessitates the use of surveillance cameras, and state-of-the-art wireless, compact, and relatively inexpensive models will offer distinct advantages over older technologies. These are to be mounted on tree trunks or other suitable uprights, and pointed at the ground. Cameras should be positioned such that their field of vision includes

the largest area possible for birds to be observed landing on the ground, allowing still for birds to be identifiable to species.

6. Study participants may wish to arrange to have captured images relayed to one's work computer. Alternatively, study participants can visit monitored locations periodically to replace camera memory cards (that may be filled to capacity with captured images) with new/blank memory cards. The minimum inconveniences of site visits can be lessened through tweaking the camera arrangement such that non-bird events (such as falling leaves and shifting shadows) do not regularly trigger images being photographed.

7. It is understood that limitations in the visual acuity of any camera system employed may not allow for captured images showing actual pellet-ingestion events. The willing researcher is reminded of the primary intent of the camera work, namely to document the number of bird landings that occur. A conservative interpretation of bird-landing data at appropriately camera-monitored sites will be that each occasion of ground surface contact equates with a pellet-ingestion event.

8. It is understood that it will be virtually impossible to identify different birds of a given species alighting on the ground at camera-monitored sites. A conservative interpretation here, as well, can be that individually captured pictures reflect a different species representative each time.

9. Given the anticipated relatively small field of vision of the cameras (perhaps 40 feet by 40 feet at best), multiple plots satisfying the provided site-selection criteria should be used. The relative affordability of dexterous cameras should allow for monitoring at least three separate plots at each of two highly pelleted locations within a county or region of a state. By way of example, recorded data of the fall zones of two different trap/skeet ranges within a county or state, with monitoring at two or three locations within each range, would sufficiently complete the task.

10. Motion-triggered images should ideally be monitored over the course of a full year or at least for as many months as grit-ingesting birds are active in the region (i.e., other than at the times when birds are overwintering at remote locations).

Study Outcomes and Applications Thereof

Potentially, the collected data may reveal that the pellet ingestion pathway is not common enough such that the phenomenon need be a focal point of ERA investigations any longer. Should bird-landing visits be found to be particularly rare, and especially where multiple study applications render such findings, the data will inform that pellet-sweeping negotiations of land owners (e.g., the military) and regulatory bodies need not occur.

Though environmentalists may insist on pellet removal actions nonetheless, documentation should capture the true basis for these. It could be that concerns over pellet-lead leaching to lower soil strata and ultimately reaching shallow groundwater provides the basis for the insistence by some that pellet-sweeping actions proceed. Should a decidedly large database be assembled that fails to validate the supposed pellet uptake pathway of birds, policy statements documenting the newly gained understanding should be scripted. Additionally, an ongoing pellet-dosing study interest would be recognized to be absent. The latter two points could, then, lead to an awareness that resources can and should be directed to ecological investigations of more weighty matters. Where pellet-dosing studies would continue, it should be made clear that only pure as opposed to applied science was being researched.

Study #6

Health-Screen Birds for Assumed Lead Pellet Ingestion; and Provide Follow-On Care

Premise

With regard to the expressed concern of environmentalists that ground-exposed lead shot pellets pose health risks to grit-ingesting upland birds, it is hoped that ERA practitioners are fluent with the limitations of certain research approaches. It should be clear, then, that conducting ongoing pellet-dosing studies can only lead but so far in efforts to secure improved understanding of ERA implications for the supposed exposure pathway. There is only so much that we can learn from the studies, particularly when the test setting is so unnatural. It should be clear, as well, that modeling efforts do not generate the information needed to know if bird populations are to be impacted. Well-intended as the models may be, their usage has only served to deflect attention away from the operative question. In cart-before-the-horse fashion, the models have been designed to generate estimates of safe (exposed) pellet-in-soil concentrations, while methods to determine if birds first bear unacceptable risks from habitat-available pellets do not exist. With these realities in mind, Study #5 discusses a novel approach to be employed in the quest to learn of the magnitude of the lead-as-grit ingestion concern, assuming that it is a legitimate one. Here, a second and related field-based study is described. It involves no artificial animal manipulation to ensure that pellet ingestion occurs, and, as with Study #5, it is purely observational in design. More specifically, this study strives for authenticating the occurrence of pellet-ingestion events, supplying quantification to the behavior where it can. More than this, the subject study encroaches on new scientific territory, for it is designed to monitor bird health in the aftermath of verified pellet-ingestion events that occur in the wild.

The study is intended to collect the necessary field information to allow for validation of the incidental shot pellet-ingestion pathway. The study is an active call to interested ERA practitioners and ornithologists to employ a biased sampling design to increase the chances of happening upon potentially health-challenged birds of worst-case scenarios. The field participant is encouraged to use his/her best thinking to secure the birds in an effort to substantiate allegations that untold bird losses are regularly occurring because myriad locations are not being swept of their pellet accumulations. As with a good many other studies in this compilation, a valid assessment of the interest can only follow from the gathered information of multiple sources, particularly where the broader geography is considered. Information brought forward from but a singular study location will not shed any light at all on the study question.

Study Guidelines

1. Identify locales where chances are very good that birds will ingest spent shot pellets of sizes 7.5, 8, or 9. Study-viable locales are those where pellets are highly exposed and will be so over the course of the year. (Areas that may initially appear conducive to study, but that will develop thick and extensive vegetation at a later period, should be avoided; the vegetation will vastly reduce the chances for birds to ingest a pellet. Also, areas that are periodically cleared of vegetation are to be avoided. Surficial pellets at these areas often become buried by the heavy mowing equipment used, again reducing the likelihood of pellet-ingestion events.)

2. Observe bird activity at prospective study sites. Specifically observe that birds at sites spend some time on the ground. A fully suitable site will be one where birds of species that are known to be grit ingesters regularly occur. Strive to secure study sites where more than one grit-ingesting species occurs.

3. For the bird netting and subsequent animal handling described in guidelines 4–8, secure an approved animal-use protocol from a relevant institutional animal care and use committee (IACUC). Note that guideline 8 will necessitate the services of an attending veterinarian.

4. Arrange to deploy mist nets at specific pelleted site locations (i.e., trap and skeet range fall zones, hunting areas) or in their near vicinity, where study team members have documented ground (foraging) activity by grit ingesters. The objective will be to trap birds with the best chance of demonstrating that they have ingested at least one shot pellet.

5. Arrange to have radiograph equipment available on demand, since birds might be netted at variable times of the day. Radiograph equipment may be arranged indoors or made field transportable to more seamlessly

advance from the mist-netting to the animal assessment described in the next guideline.

6. Carefully hand-transfer netted grit-ingesting birds to individual coni-cal plastic restraint devices. (These can simply be clear plastic 24-ounce soda bottles that have had their tops and bottoms removed.) Restraining devices should be taped to the radiography image-detector plate to restrict device movement. Left-right lateral projections are to be sought. Note: appropriate radiation safety procedures for personnel involved, with respect to duration of radiograph equipment usage, body shielding, and appropriate distancing from the equipment, must be implemented.

7. Maintain a log of all radiographed birds, specifically recording species, sex, age, and the number of shot pellets observed.

8. Retain all birds with pellets observed via radiograpohy in the laboratory, singly housed, and supplied with water and appropriate feed *ad libitum*. Minimally monitor birds until pellets have either been excreted (verifi-able by analyzing excreta in cage floor droppings) or have completely dissolved (verifiable with radiography). Specifically record general appearance, posturing, activity levels, and survival. Note: researchers should recall that ingested shot pellets do not behave at all like true grit particles that are retained in the ventriculus. If not excreted in the shorter time frame, maximum pellet retention times will be of approxi-mately 3 weeks (for, in that time, pellets will fully dissolve). To the extent practicable, retain birds for several weeks post pellet excretion or pellet dissolution. Upon completion of the monitoring phase, release birds to the field at the point of capture.

Study Outcomes and Applications Thereof

Through its arrangement, the study places squarely on the researchers' shoulders the onus of demonstrating that incidental pellet ingestion by upland birds poses a severe environmental problem. Part 1 of the step-wise design is to procure birds that may be health-challenged. Where repeated efforts result in few (or no) birds with ventricular pellets, researchers will need to be answerable for that outcome, for it was they who applied their best judgment (e.g., in the realm of site selection) to ensure they would arrive at some. Opining that finding birds bearing one or more ventricu-lar pellets is simply too difficult a task will not be an acceptable response. Should few netted birds bear pellets, it will also not be passable as a defense to argue that all birds that have ingested pellets have died, with the netted birds soon to arrange their own fate. Researchers should con-sider if netting at bird-hunting sites would yield more pellet-laden birds than at firing range fall zones. Perhaps this could be the case, although the former are unlikely to bear the characteristic focused and highly visible accumulations of the latter. Numerous ecologically relevant pellet-dosing studies have demonstrated that birds can well tolerate as many as three pellets administered at the same time, a situation quite unlikely to occur in nature. The findings to emerge from the monitored indoor-housed birds can potentially go a long way towards validating the concerns of environmentalists (i.e., that birds with ingested pellets are fated to die), or demonstrating that the pellet exposures are inconsequential. Where birds with ingested pellets are only rarely observed, and where birds with pellets survive through to the time that pellets are either excreted or dissolved, and beyond, adjusted site-management strategies can follow. The intent to sweep sites of accumulated pellets raises issues of habitat disturbance, to include the accidental damaging of trees and/or the deliberate removal of trees along with other site features. Potentially, concerns over substantial lead pellet accumulations imposing physical stress to site biota can be dispensed with, because the information collected in support of this study will indicate that pellets can be left in place.

Study #7

Apply a "Top-Down" Medical Screening Scheme for Birds and Mammals

Premise

Physicians are often able to diagnose medical issues in their patients by reviewing the results of standard tests such as routinely run blood analyses. Knowledge of the normal range or set point for a given parameter (e.g., a specific hormone measured in blood) allows for diagnoses to proceed. A physician, therefore, can diagnose hyperglycemia (and perhaps diabetes mellitus, as well) in a fasting subject, if the patient's blood glucose level (perhaps 230 mg/dL) exceeds the established normal range for it (of 70 to 100 mg/dL). In the same way, ecological receptors can be health-assessed by comparing their analyzed blood and urine profiles against established profiles for these fluids that depict the norms. This study, then, is necessarily a linked follow-on endeavor to Study #15. That is, it would otherwise be premature to collect biological fluid samples from receptors at contaminated sites with the intent of analyzing and characterizing a group of metabolic parameters, before establishing that certain parameters are amenable to review and quantification altogether. Consider that for a given parameter, the yield (volume) of a sex hormone after sample preparation could be too low to allow for reliable measurement. That we should sooner learn of such a study impediment from animals culled from pristine locations than from animals culled from contaminated ones would make for a prudent measure.

Primarily, this study asks motivated researchers to systematically collect the blood (and urine samples, if appropriate) of contaminated site mammals and birds in support of a biological fluid-based "top-down" health assessment scheme. The scheme's essential task is one of documenting exceedances of established species-specific normal metabolic values (such as are discussed above), since such exceedances are indicators of site (chemical) stressors having offset homeostasis. Where x-rays, blood pressure readings or other non-biological fluid-based measurements differ notably from those maintained in a repository of information for animals of pristine locations, these too are to be documented.

Study Guidelines

1. After seeking IACUC approvals, trap mammals and birds at established contaminated sites, briefly restraining them to facilitate venous blood collection. Endeavor to collect the samples from sites where HQs have been computed for the species being evaluated. Ensure that the relevant soil COC list is collected from each site, and that the chemicals considered to be risk drivers are also recorded. (Note: No soil samples are to be collected as part of this study. Qualifying study sites are those that have their soil contaminant profiles already known.)

2. Endeavor to sample as many species as possible, and to sample at least five individuals of each species. Where the database supporting the metabolic parameter comparison effort reveals distinct differences in parameter measure magnitude between sexes, endeavor to sample five members of each sex.

3. Small rodents should necessarily be collected in all instances. These animals, it should be noted, need not be euthanized to enable blood collection, for venous blood can be collected from the end of the tail.

4. Compare individual and mean parameter measures in blood against database values. Endeavor to compare measures from contaminated sites against database measurements for the locale that is geographically closest to the contaminated site. (Thus in most instances, state-specific database measures should be used where these exist.) To the extent that measurement variation is minimal across the database for a given species, site specimens can have their measurements compared against the database mean.

5. For all instances of site animal measures in blood or urine that are outside the normal value range, tabularize the site animal's observed general health condition, the respective contaminated site's risk-driving chemical(s), the animal's age, and the likely maximum age the species is expected to attain in the field. Do the same for all instances where non-biological fluid measures (chest x-ray) deviated from the database/repository norm.

6. Tabularize instances of blood and urine parameter measures that exceeded the normal range, along with respective NOAEL- and LOAEL-based HQs that were below 1. Separately tabularize instances of parameter ranges that were not being exceeded, but where respective NOAEL- and LOAEL-based HQs were greater than 1. Endeavor to explain any non-aligned information (i.e., instances of parameter

exceedances and safe HQs; absence of parameter exceedances where HQs fail).

7. Assemble a comprehensive analysis of the ecological significance of exceeded metabolic parameters. Specifically include older animals in the analysis (i.e., animals that are near the end of their lifespan, as in a fox estimated to be about three years old). Specifically include small rodents in the analysis, particularly because these will almost never live beyond a year, and because their presence at multiple-decade-old contaminated sites necessarily demonstrates that offset homeostasis, should it be observed, is not problematic for them.

Study Outcomes and Applications Thereof

This study, quite obviously, is specifically intended to heighten the awareness of ERA practitioners, whose task it is to assign determinations of chemical stressor-imposed impact, should they believe they have observed such. Where contaminated site receptors present with metabolic parameter measures that exceed the normal range, ERA practitioners and risk managers will be challenged to conclude that site receptors are at risk or impacted. This follows, in part, from the unparalleled nature of the collected data, i.e., never before will blood lipid profiles, for example, have been analyzed in birds and mammals of contaminated sites. Another critical unknown is the degree to which metabolic (or physical) measures need deviate from the norm to constitute a health effect of note. While the utility of this study and the preceding one may appear to be questionable due to thresholds-for-effect not being defined, this is anything but the case. Venturing into the study's comparative analysis (where exceedances might be highlighted) will necessarily cause ERA practitioners to appreciate that labeling differences (here, deviations from the normal range) as adverse, is without basis.

Where metabolic parameter measures exceed the normal range, ERA professionals will have pause to consider how long the measures have been deviating from the norm (it being wholly unrealistic to think that the deviations only coincidentally arose in the year that the site animal measurements were collected). These professionals, too, will be forced to consider what utility, if any, lies in observed exceedances where ecological receptors are senescent (e.g., a mouse or a rat estimated to be about a year old), and having already contributed in a reproductive capacity.

The seemingly more straightforward (and anticipated?) outcome is that there would be no parameter measures exceeding the normal range.

Although not readily testable, several phenomena could account for this outcome. These would include metabolic measures being tightly buffered in the wild-type animal, decades passing, during which time exposed animals come to adapt to a contaminated site condition, and site contaminants not being of sufficiently high concentration and/or not being sufficiently bioavailable to allow for metabolic measures being thrown askew. Absent metabolic parameter exceedances in site-exposed animals can easily support determinations of sites affording adequate protection to ecological resources. That is, there should be no reason to suspect health impacts in site receptors where their metabolic profiles are in line with the norm, even though threshold-for-effect information may not exist.

Study #8

Establish an Environmental Residue Effects Database for Terrestrial Animals in the Wild

Premise

Some two decades ago, the US Army Corps of Engineers (ACOE) introduced a database to expand ERA capabilities for the aquatic arena. Updated numerous times since being first made available as a toxicity assessment tool, it sought to relate chemical concentrations in the tissues of aquatic species to the health effects the specimens bearing those chemical loads would display. Unfortunately the intent and utility of the effort never really materialized. While the database swelled with the information of more than 700 studies, and reported on close to 200 different species and more than 200 different analytes, it was not possible to learn of the tissue concentrations that would impede the health of aquatic ecological receptors as we would encounter them at contaminated sites. Much of the database reported spiked water column concentrations and the tissue concentrations of introduced (fish and other) test species exposed in the laboratory. A good deal of the database reported on the tissue concentrations measured in aquatic species drawn from natural waterbodies with contamination histories. Most critical to note is that effects reporting was almost always absent from the gathered information; assimilated tissue concentrations are not "effects," something that ERA practitioners so commonly forget. The reduced or non-existent utility of the database, touted at one time as housing some 3500 results, was clearly recognizable. While we might know how a chemical manifests in a fish (e.g., the quantities and concentrations that assimilate; the differential partitioning of chemicals throughout the body), we won't know if a chemically burdened fish is constrained in some way. Potentially, chemically burdened aquatic species fail to achieve their proper size, locomote erratically, behave abnormally, or reproduce in inferior fashion. Advertised nevertheless as a database that relays "apparent effects," the ACOE-developed product perpetuated the misconception that the database was truly serviceable. Not knowingly, the database also forestalled far more exacting aquatic toxicity research from coming about. The proof for this is the continuing ERA practice of the reporting of, say, whole-body chemical concentrations, and concluding based on these, that site monitoring or site remediation is absolutely necessary. We can profile further the misapplication of the science. If we call an ERA practitioner on his/her steadfast allegiance to the so-called "apparent effects," we are likely to hear a scripted retort, complete with an ungrounded scientific basis, about health effects in other aquatic or piscivorous species: "The fact that we detected contaminants in fish tissue means that every other species in the food-chain is also being exposed to the contaminants, and we cannot stand idly by allowing for continued chemical exposures to occur." With such a retort, we are once again back to guilt-by-association. More importantly, the effects over which the well-meaning ERA practitioner is so concerned, and that should have already cropped up years ago, will not have been described.

Presumably chemical-in-tissue concentrations for any species, terrestrial or aquatic, are biomarkers of health effects. Why researchers to this point have merely reported associations of chemical concentrations in media and body tissues, and forgone the next step of establishing those chemical-in-tissue concentrations that signify impacts that disturb the ecology, should be viewed as a mystery. The field appears to have taken the easy way out and, not knowingly, the regulatory community might be responsible for the absent information. With regulators content to cite chemical uptake into tissue alone as a sufficient basis to advocate for cleanup work, why would the masses bother with doing any more? This study, quite broad in scope, allows interested parties to demonstrably move ERA science ahead, facilitating an understanding of when assimilated chemicals make for legitimate ERA concerns and when they do not. As with a good number of studies in this compilation, there is more than ample (perhaps *unlimited* is a more accurate term) opportunity for interested parties to participate. As with a good number of the studies too, where independently acting researchers can collaborate to aggregate the information they have secured, the benefits for the ERA field can be the most constructive.

Study Guidelines—Field

1. There are only limited laboratory tasks for this first (field-based) portion of the study. They include conducting chemical constituents analyses on certain animal tissue samples collected in the field (e.g., blood, fur; see guideline 8), and analyzing chemical profiles in soil, or possibly, too, certain dietary items. Animal dosing to either learn of manifested tissue concentrations, or in order to document health effects that might arise pursuant to an animal's chemical dosing, is reserved for the study's second portion. The researcher is reminded that (a) this study's true interest lies with understanding the health effects, if any, of animals chemically exposed in their natural settings, and (b) animals in the wild substantially differ, in a toxicological sense, from animals that are experimentally dosed/artificially exposed.

2. Select study sites that are sufficiently large to allow for the collection of enough animals of a given type (reasonably five) to, in turn, allow for reasonable study findings to emerge (i.e., such that trends can be observed, and that data compilation will be statistically sound). By way of review with regard to mammals, one-acre sites and the like would only be serviceable for small rodents, given the notably reduced home ranges of these forms. Review the relevant literature then, for the densities (in units of animals per hectare, acre, or square mile) and home ranges of the species that one might wish to study. Red fox (*Vulpes vulpes*) could not be reviewed for the subject study unless a contaminated terrestrial site was of at least 168 acres (the area that, based on maximum density for the species, would reliably hold five animals). Long-tailed weasel (*Mustela frenata*) would also not be workable for a great many contaminated sites in the United States, given this species having densities of 16 to 18 per square mile (640 acres; 40 to 47/km^2) and home ranges of 30 to 40 acres (12 to 16 ha). Would-be sites of interest, unless they were of some 300 acres, would not supply the requisite number of weasels needed for study, and for (smaller) more common sites, a given weasel would only be allocating mere fractions of its time at these. Further, while the intent would not be to spend the captured animals, it would not be ethically correct to handle the exhaustive population of a smaller area, given that these animals might be harmed or offset in some way through direct/hands-on study.

3. Select one or more mammalian or avian species of choice to study at contaminated terrestrial sites. Given the variability of contaminant suites at sites and the ranges of assumed-to-be ecology-compromising chemical concentrations, collect surface soil samples (and soil samples of other appropriate depths as needed) to be analyzed, unless useable soil concentration information already exists. If there is any question that "site" soils might not be chemically contaminated, analyze the soils *before* proceeding with the animal work (this in an effort to ensure that animals are not being culled unnecessarily).

4. Secure all necessary IACUC permissions before conducting any animal trapping and handling.

5. Record the body weights and overall health conditions of the mammals and birds sought. Record somatic measurements (e.g., foot length, tail length, wing span) that might differ from those of reference location animals only if it is known how much difference in measure there need be in order to signify a compromised biological function. Recall that the study is not interested in observing if physical size measurements have been influenced by chemical exposures, but rather in knowing if/ that animal behavior or physical abilities of site receptors have been compromised.

6. Where possible, develop means for evaluating behavior, locomotion, feeding, and reproduction in the field. Such will require the design of stress-free enclosures that allow (relatively) unobstructed observation of habits (e.g., climbing, feeding, preening, nesting), and for recording ambulatory or flying speed, and, in the case of birds, take-off and alighting abilities, etc.

7. Identify all instances where site animals display less-than-optimal performance for any function or activity. If observing reduced performance would only be possible through a comparison with animals that reside in contaminant-free (and, of course, nearby habitat-matched) reference locations, subject animals at the latter to the same testing and observation protocols applied for contaminated site animals.

8. Only where compromised performance is observed (e.g., a species does not ambulate normally; a species is noticeably moribund), should (blood, fur, feather, etc.) samples be collected and submitted to a laboratory for tissue residue analysis. If the study animals are those that can be harvested from the field (e.g., small rodents), have some specimens submit to whole-body analysis, and have the major body organs, blood, and fur chemically analyzed in the others.

9. Align all instances of observed compromised performance (such as are mentioned in guidelines 6 and 8) with the presumed assimilated chemicals responsible, and endeavor to align the degree of compromised performance with the measured tissue concentrations. Where multiple chemical stressors are present at a site, endeavor, to the extent possible,

to identify the chemical(s) principally responsible for the compromised performance.

10. Endeavor to assess reproductive effects, recognizing that having removed animals to artificial enclosures may clearly interfere with reproduction.

Study Guidelines—Laboratory

1. As with the field-based portion of the overall study (guideline 4), secure all necessary IACUC permissions before commencing work.

2. Testing one chemical at a time, expose birds and mammals (likely, only mice or rats) by spiking the diet and drinking water (simulating a natural means of contaminant availability and uptake) with the intent of achieving measureable and biologically significant tissue concentrations. Ensure that the artificially arranged dietary item and drinking water chemical concentrations are reflective of those that might appear in the wild. (Note: this portion of the study is not a toxicological one, where the primary interest is in learning of the capacities of certain chemicals to cause effects, and where deliberately ignored is the circumstance that applied doses exceed those that would actually present themselves in contaminated site settings.)

3. After allowing for a reasonable exposure period (perhaps 3–4 months), apply guidelines 5 through 7 of the study's field-based portion. If none of the earlier-listed performance features are offset in the dosed test animals, terminate the study; for such an outcome, there is no purpose in learning how diet- and drink-supplied chemicals manifest in blood and other tissues.

4. Apply guidelines 9 and 10 of the overall study's field portion. Note that rodent reproduction can be skillfully assessed with the patented Rodent Sperm Analysis method (U.S. APHC, 2009; Tannenbaum and Beasley, 2016), recognizing that the method necessitates a study control group. As with the preceding guideline, only if (rodent) reproduction is found to be compromised, evident in any of the RSA method's thresholds having been exceeded, should chemical tissue residue analysis proceed.

Study Outcomes and Applications Thereof

The thinking behind constructing an environmental residue effects database is fair enough. Chemical concentrations in certain tissues might be indicative of critical biological effects that have taken hold in examined animals. (All too often, ERA practitioners are wont to think that "effects" spoken of in an "environmental residue effects database" refers to possible impacts that might arise in animals that have consumed others that bore a distinct tissue burden, but such is not the case.) Presumably, the ACOE-headed initiative began with the proper intent, but no one ever bothered to relate such things as a fish's liver concentration of a given chemical with its swimming speed, the fillet concentration with its overall growth, or the kidney concentration with its reproductive success. Reporting the chemical concentrations that actually registered in either whole-body fish placed into spiked aquaria, or in (whole-body) fish drawn from contaminated waterbodies only tell us loosely of chemical uptake. But is there anything wrong with fish that bear a tissue burden? It goes without saying that the preceding paragraphs describing efforts to be taken for birds and mammals most definitely need to be applied also to fish, with the intent of shoring up a broad and long-standing data gap. We should not always be suspect of health effects accruing to fish because they bear chemical residues, when we could know rather definitively that a chemical residue—the reader should please excuse the pun—may amount to nothing more than a red herring. With assembling an environmental residue effects database for any animal, the intent is to not only ward off the uninformative and misleading food-chain modeling that continues to populate ERAs. It is to inform on whether ecological receptors themselves are health-compromised due to chemicals they may accumulate. Recognizing that sites to be investigated have been contaminated for decades by the time they submit to ERA should be sobering in the study context. Hence, the described study directing enthusiasts to document abnormal health and/or behavior before delving into the prospect of analyzing tissues. If evidence of the former should be lacking, there'd be no point to establishing the tissue burdens of observed and/or tested animals.

As with several other studies in this compilation, it is a fair anticipation that researchers will discover that contaminated sites are simply too small to allow for capturing a sufficiency of animals to evaluate. Such constitutes a tremendous side-benefit, or perhaps the main benefit, in taking on the assignment. ERA is altogether unnecessarily investing energies in documenting tissue burdens, when there are too few animals roaming a property to matter. For the terrestrial component of ERA, the absence of true effects (and not so-termed "apparent" ones), defined as animals being limited from doing something, or from doing it so well because of tissue-assimilated chemicals, can speak reams. Absent effects should mean that ERA can forever dispense with body burden concerns. In the larger picture, where altered functions and behaviors are not demonstrated, we will again learn that animals are more chemically resilient than we realize, and we stand to discover (again) that there may be no need for ERA.

References

Tannenbaum, L.V., Beasley, J. 2016. Validating mammalian resistance to stressor-mediated reproductive impact using rodent sperm analysis. *Ecotoxicology* 25:584–593.

U.S. Army Public Health Command 2009. Rodent Sperm Analysis. U.S. Patent 7,627,434.

Study #9

Aggressively Expand the RSA Database

Premise

RSA remains the only direct health status assessment method in ERA. This vetted and patented method is predicated upon thinking that seems to evade the mind-reach of ERA practitioners. This is somewhat hard to understand because the method's fundamental underlying premise is simple enough to follow, and certainly does not enter into the ranks of the proverbial "rocket science" about which many jest. RSA posits that there is a statute of limitations, so to say, for health impacts to first arise in ecological receptors contacting contaminated sites. Given the ages of contaminated sites by the time they submit to ERA, and the very brief generation times of ecological receptors, if effects are not apparent today, they are never going to crop up. The quintessential question, then, asking of the potential for an ecological receptor to develop a toxicological endpoint at a site that has been contaminated for multiple decades, is without backing and, to be fair, quite ludicrous. If impacts were ever to arise in site receptors, especially those that have no liberties whatsoever to evade their contaminated environs (owing to their non-migratory nature and having miniscule home ranges), we would know of these already. Contaminated sites would be marked by the conspicuous absence of biological forms well-known to us. Assuming site contamination hadn't erased their existence, at the very least, animals would present with overt signs of illness. It is clear then, or it should be clear, that the rightful assessment approach to adopt is one of looking to see if ecological health impacts *have been* elicited. And of course, from here on, we should dispense entirely with what have only been feeble and hopelessly failing machinations to predict impacts. For the present, it is disheartening to see ERA practitioners consumed by the HQ-based approach to assessment, and to a point where they would never think to pull back from it in order to try something else.

Something to consider: think of a contaminated site with a chemical release history dating back to the Korean War. Perhaps the site is a former TNT manufacturing plant, where, during the height of production to support the war effort, many spillages occurred, and vast quantities of off-spec explosives were deposited, uncontained, on the ground. At the time of this writing, then, explosives contamination in soil is some 65 years old, with the site already far too old to be submitting to HQ-based ERA. The site, nevertheless, is in the queue for having an ERA conducted. The site becomes a bit of a political football, with different state and federal entities vying for the designation of lead agent, and consequently it sits a few more years with this issue unresolved. A new administration comes in and, in the reshuffling of environmental stewardship authorities, all tracking of our dear site is shelved. We fast-forward to the year 2050; our dear site is 100 years old when the administration then in place introduces an initiative to, once-and-for-all, close out from earlier environmental programs any sites that proverbially slipped through the cracks, such as our Korean War-era explosives site. Our long-forgotten site is rediscovered and slated for an ERA, to be completed in 2051. The likelihood is high that the HQ-based assessment scheme we presently have will still be in vogue at that time. We would be correct to anticipate the ERA to be assembled, describing at its start having the purpose of determining the potential for site ecological receptors to be at risk. Would any ERA practitioner at that time, to include the ones tasked with producing that ERA, honestly believe that their work might uncover *a potential* for ecological impacts to set in? The question is asked because the hypothetical scenario is not all that far-fetched. Presently there are no indications that movers and shakers in the ERA field are at work developing methods akin to RSA, with a focus of documenting what has happened. Further, no one is engaging in such work because movers and shakers see no need to move beyond HQ-based assessment. (Being a mover or a shaker might, at best, connote trying to bring online HQ-based assessment for the inhalation or dermal contact chemical uptake routes.) The proof of this is clear; ERA uncertainty sections never acknowledge a temporal statute of limitations for site ecological receptors to be toxicologically undone by the sites at which they reside.

This study, perhaps the most straightforward of those in this compendium, invites inquisitive individuals to immerse themselves in RSA application, seeking out opportunities to expand the database of reported outcomes. With only two criteria to be satisfied for method application (namely, that sites be verified as bearing contamination and supporting rodents), opportunities to partake in the work ahead are really limitless. Additional applications of the method will necessarily secure information of a broader range of sites (in terms of ecosystem types, contaminant suites, and site histories), and importantly will expand the list of rodents species reviewed.

Study Guidelines

1. It should be clear to interested parties that the long-range objective in greatly expanding the universe of RSA sites is not for method verification (or even for simple method tweaking); there is no question that the RSA method is fully developed and matured. The objective, rather, is to allow for high-order quantification of the method's essential reporting outcome and *raison d'être*, namely the definitive determination of site (soil) contamination having impacted mammal reproduction or not. Truthfully, there are other outcome trends to also track, if not codify, should these quantitatively be found to be clear-cut. Thus, trends in sperm parameters (e.g., for count, morphology) at contaminated sites relative to reference locations, those for a greater range of (rodent) species than have been reviewed to date, and those for a broader geographical range than has been thus far studied, are of interest at this time. Of course, there is great interest in knowing of specific chemicals or chemical mixtures that make for sperm parameter shifts, and the concentration ranges that trigger them. (See "Study outcomes and applications thereof.")

2. For sites that are to submit to RSA, an effort should be made to document how long each has been contaminated.

3. Where possible, sites should be sought out that have had ERAs conducted, and where unacceptable HQs resulted for rodents and other mammals. It is imperative that RSA not be run at sites that have had soil remediation of any sort, to include having a layer of clean topsoil applied.

4. The essential RSA method is provided here. It is suggested that researchers avail themselves to the RSA method technical guide, and to the group of published articles on RSA appearing in peer-reviewed journals.

5. Identify one and preferably two potential reference locations for each viable contaminated site. (A back-up reference location anticipates the possibility that an intended reference location is found to not be suitable for one reason or another.) Recall that suitable reference locations are habitat-matched to contaminated sites, and are located relatively close to them, but not so close such that a given rodent could appear at both.

6. Establish the rodent species to potentially submit to RSA. Ensure that appropriate state and other animal trapping permits, and appropriate IACUC approvals, are secured before commencing any work. Note that the method cannot be applied unless there is at least one small rodent species co-occurring at a site and its paired reference location. Note additionally that multi-species comparisons embellish the level of information RSA applications provide. See next guideline.

7. As a rule, the RSA method cautions against working with shrews (e.g., *Blarina brevicauda*), recognizing that these animals, with their excessively high metabolism, will often be found dead in morning-checked traps. In the spirit of gleaning the most RSA information, though, it is recommended that an effort be made to trap enough shrews of a species commonly occurring at a site and matched reference location to support a valid method application. So long as animals are not dead when traps are checked, sperm count and morphology (but certainly not sperm motility) should be unaffected. Best professional judgment should be applied to decide if shrew trapping at a given site should proceed if it is known *a priori* that shrews are the only small rodents (technically not rodents, but insectivores) available for RSA study.

8. Within an approximate 7–10 day period, conduct rodent-trapping at both a site and its paired reference location. Saturate sites and reference locations with spring-loaded and appropriately baited safety traps, with the goal of collecting 15 adult males of as many species as is feasible. For all captures, to include juveniles and females, record weights and overall health conditions. Assess females as to being pregnant or lactating. Note that RSA may be workable with as few as 8 or 9 "keepers" (i.e., adult males) in lieu of the idealized count of 15.

9. At a minimum, assess sperm count and sperm morphology following descriptions in the method's technical guide and/or the peer-reviewed literature. Note that, while assessing sperm motility at paired sites is most desirable, it is recognized that tracking this parameter may be budget-prohibitive.

10. Recall that in RSA's conservative assessment scheme, sperm parameter population means for contaminated sites that are shifted unfavorably from the reference location arrangement (i.e., lesser count, a greater incidence of morphological deformities, lesser motility) need not be statistically significant. Sperm parameter data are screened against the nominal thresholds-for-effect of a 60% count decrease, a 40% motility decrease, and a 4% increase in morphological anomalies.

11. Identify all sites that had one or more sperm parameter thresholds exceeded, tabulating the degree of exceedance, the site chemicals that might be responsible for the outcome, and, in that case where an ERA was previously conducted, the site rodent HQs that were computed.

Study Outcomes and Applications Thereof

None of us have a crystal ball available that can foretell outcomes of the sensitive direct health status assessment method for mammals that is RSA. We are therefore free to anticipate the outcomes of what hopefully will be tens and scores of additional RSA applications to happen. Earlier in the premise, the study's straightforward nature was mentioned. More than simply alluding to the novelty of the existence of some outcome information (albeit not as much as would be preferred), it was alluding to what appears to be an already-demonstrated trend, or, more correctly, a small series of trends. First, there have been no instances of sperm parameter threshold exceedance observed (where count, motility, and morphology have all been monitored). Second, the count is almost always (nominally) lowered at the contaminated sites, but not lowered enough (of course) to trigger a conclusion of compromised reproduction. Third, in numerous instances, the motility has been (nominally) higher at the contaminated sites relative to the reference location condition, with this suggested to be an earmark of a certain compensatory effect occurring, as follows. Since a lower count can (at some point) serve as an impediment to reproductive success, a bolstered motility score (i.e., an increase in the percentage of properly swimming sperm relative to the percentage of proper swimmers in reference location samples) can only be a reproductive success asset. In that these trends or quasi-trends have arisen for rodents assessed at sites with highly variable patterns of contamination, in widely divergent habitants of distinct geographic regions, and for more than half-a-dozen species, the proverbial writing may be on the wall. It might simply be that mammal reproduction, as viewed through the lens of responses of rodents that all share an intimate contact with affected soils, cannot be offset in the actual contaminated site setting. We cannot guess at this, and thereby the value that lies in the data yet to be acquired is both acknowledged and great. Those truly fluent with RSA theory and those who recognize that instances of ecological impact having occurred at chemically contaminated terrestrial sites are virtually unknown already understand that the intended work is so very unlikely to identify instances of sperm parameter threshold exceedance. The intent of greatly augmenting the existing repository of RSA outcomes, then, is to raise consciousness. Recalcitrant ERA practitioners can come to shed their bias that takes the form of an insistence that ecological impacts abound at sites, when a wealth of information spanning different types of sites, and reporting on multiple species, indicates otherwise. Where an appreciation is secured that ecological receptors in the wild are more resilient to chemical stressors than they've been given credit for, the door is left open for instituting a shift in ecological assessment gears. Should a great many, if not all, RSA outcomes indicate that reproductive impacts are absent at contaminated sites—a reasonable finding—ERA practitioners, to include regulators, can be expected to latch onto the development of other direct health status assessment methods for other mammals, and a number of other animal groups occupying terrestrial and aquatic ecosystems.

Study #10
Reposition Small Rodents in the Wild to Enhance the Understanding of Contaminated Soil Exposures

Premise

With great deliberation in previous writings over close to two decades, the author has placed an extreme emphasis on the valued information that certain site receptors stand to convey to the attendant ear of the dedicated ERA practitioner. In a nutshell, for those who are willing to learn from what chemically exposed animals have to show us—as opposed to those whose only interest is in replaying, *ad nauseam*, an uninformative formatted ERA process—contaminated sites should be seen as living laboratories. Unbeknownst to those responsible for having left contamination footprints behind, when each chemical release occurred, a dosing study of sorts was being initiated. Nature did all the subsequent work; there was no need for anyone to purchase or house animals to serve as test subjects, to artificially adjust animal diets, or to manipulate any other aspect of an animal's existence. There were no light-dark cycles or thermal regimes that had to be rigidly maintained, if not fixed. (The only aspect of the ambient condition that is fixed is its variability!) This analogy to conventional (indoor) laboratory animal dosing studies continues: when we happen upon contaminated sites that are to submit to ERA, the chemical "dosing phase" (although it might still be ongoing) is, for our needs, complete. The salient point, then, is that contaminated sites provide the great luxury of letting us know decidedly how ecological receptors fare after decades of chemical exposure and tens or perhaps hundreds of generations having been chemically exposed. The duration of conventional laboratory study exposures pales in comparison to the arrangement in the wild. Two other related notions have been conveyed in the context of the laboratory study analogy. First, with so much time elapsing at sites by the time ERAs are done, ecological impacts (should there be any to observe at all) are today

71

at their height. Second, the concept of setting to work in the present day to assess risk is laughable; after so much time elapsing, we cannot honestly say that we are interested in learning of a "potential" for health effects to arise.

There is no doubt that the small rodent holds the key to a bettered understanding of ecological effects at terrestrial sites, and in particular securing either the point or the pointlessness of involving ourselves with the formatted ERA process as we have it. The small rodent is the only mammal that can regularly be culled from the field, and for which obtaining IACUC permissions proceeds smoothly. In the main, the subject study does not involve culling animals from the field *per se*, but rather strategically relocating them. With sound purpose to the work, well-written study protocols to pave the way for the subject study to proceed stand to be readily accepted. The specific premise is straightforward enough: to collect small rodents of non-contaminated areas and to place them at contaminated sites. And then the reverse: to remove small rodents from contaminated sites and situate them in appropriate contaminant-free zones. Undeniably, this study necessitates an engineering feat. Study success will revolve around the means by which repositioned animals can be effectively contained in their environments such that they can be captured for further study after a period of time.

Study Guidelines

1. Design a deployable large area enclosure of approximately 0.25 to 0.5 acres, fully capable of containing small rodents (effectively mice, rats, or voles only) that are to be placed within them for periods of approximately a half-year. Throughout the enclosure's perimeter, seal the soil-enclosure interface such that animals cannot escape (other than their burrowing anew and establishing tunnels beneath the enclosure's walls). Walls of the enclosure can be made of any material, and might best serve researchers if panels are of a see-through Plexiglas, which should not be scalable by rodents. The enclosure design should include a singular means of ingress/egress for researchers who will be conducting animal trapping (see guideline 4). Roofing for the enclosure is strongly suggested to preclude the possibility of predatory birds having access to rodents at nighttime. Where roofing is used (e.g., thick rope netting), it should not intercept a significant amount of sunlight, and should not impede precipitation.

2. Plan to conduct the work at multiple locations. These should be hazardous waste sites and the like that have yet to submit to cleanup activities.

Record the range and average concentrations of each site-defining chemical. For each study location, identify a relatively nearby, habitat-matched (non-contaminated) reference location. Perforce, the reference locations will support the same rodent species as those that occur at their paired contaminated study sites.

3. Live-trap two-dozen small rodents from a reference location, and record sex, age (as juvenile or adult), and general health condition. Tag the rodents (perhaps by ear punch) such that they will be easily recognizable upon recapture. Release the tagged rodents at varied locations within a contaminated site's deployed enclosure. A secure way to do this would entail lowering the animals over the enclosure wall at random points about the enclosure's periphery. Given the relatively brief lifespans of rodents in the wild, and the intended significant portion of the lifespan to be monitored (see next guideline), endeavor to work with juvenile rodents (and not even sub-adults, to the extent that a differentiation can be noted). Understand that if contaminant site-released rodents are too old to begin with, reduced trapping success after six months (see next guideline) could be purely age-related (i.e., independent of any chemical exposures incurred).

4. After six months, carefully enter the enclosure through its access point with all necessary trapping equipment. Ensure that the entranceway is immediately shut. Set out two grids of approximately 100 appropriately baited traps each, for one or more nights. Endeavor to trap the previously tagged (released) animals over the next few days. Endeavor to also collect 20 additional (i.e., non-tagged) rodents there; see guideline 7.

5. Record animal weights and general health conditions of the recaptures and the non-tagged animals. Where the trapping effort for recaptures is less than 75% successful, endeavor to account for the situation.

6. At the reference location, trap an additional 20 rodents of each species that was initially collected, tagged, and released at the contaminated site. In half of these animals, measure site chemical concentrations in liver, kidneys, and spleen. Run whole-body analytics for site chemicals on the remaining (10) animals. (Note: assuming the reference location is properly situated [i.e., distanced sufficiently from its paired contaminated site], there should be no expectation of detecting site xenobiotics in the tissue or whole-body samples.)

7. As with the previous guideline, run analytics for the site's chemical suite in liver, kidneys, and spleen on half of the contaminated sites (recaptured) tagged animals (10) and half of the non-tagged animals (10). Run whole-body analytics for site chemicals on the remaining (10) tagged and non-tagged site animals. Endeavor to run all the analytics (i.e., organ and whole-body samples of site and reference location animals) on the same day.

8. Endeavor to collect 20 additional small rodents from the site. Tag and relocate these to the reference location which must, at this point, have

a deployable enclosure in place. After six months, endeavor to trap and reclaim the 20 animals, and observe overall health.

9. Tabularize the analytical reporting, comparing (a) organ and whole-body concentrations of reference location animals, (b) tagged animals trapped after six months of exposure at the site, (c) site animals, and (d) site animals after having been relocated to the reference location.

Study Outcomes and Applications Thereof

For evaluating health effects to ecological receptors from contaminant exposures, the normative practice is to conduct all work in the laboratory. Under rigidly fixed ambient conditions, commercially reared organisms are placed into artificial containers (e.g., fish in aquaria; earthworms in jarred soils), unnaturally dosed (e.g., via gastric intubation or i.p. injection), or fed artificially amended/tainted diets (i.e., with test chemicals laced into feed or water) for a period of several weeks. With nary an exception, a singular chemical at a time is tested in the conventional laboratory arrangement, although it is an extremely rare instance that a contaminated site submitting to ERA investigation has but a singular released chemical of concern. While these experimental designs offer certain conveniences, the chemical exposures they explore are far removed from those that concern us (or should concern us) at contaminated sites. The limitations to the information to be brought forward through the conventional designs speak for themselves. This study recognizes that a far superior way to study the potential for site chemicals to trigger health effects in mammals takes the form of transporting the very species we assess from the true field condition to contaminated sites. And so it is that not only is the testing environment a natural one, but that the very animals that live outdoors and in proximity to contaminated sites, as opposed to commercially bred animals, are those that are necessarily employed to get at the answers being sought. The above notwithstanding, accounting for poor recapture success (at either sites or reference locations) may present a formidable challenge. It may simply be that tagged rodents can outwit the deployable enclosures to the point of escape, or that, in some unintended way, the enclosures increase the vulnerability of rodents to predation. To the extent that a multiplicity of enthused scientists engage in this study, an ultimate scientifically efficacious enclosure design will emerge—one that should be freely shared.

Findings of high recapture success for contaminant-free rodents placed at contaminated sites for six-month durations will first confirm that the deployed experimental enclosures were effective in containing the focus animals. High recapture success will also confirm that soil, plant, worm, insect, and other chemical residues at sites do not threaten survivability for small rodents serving as both maximally exposed mammals and sentinels for others. Given the great array of contaminant suites at sites, the more instances there are of recaptured animals with no apparent signs of toxicological challenge, the easier it will be to conclude that certain elements of the desktop ERA approach are really unnecessary. Primary among these would be the routine HQ exercise for one or more rodents. (Conspicuously absent documentation of depauperate rodent populations at contaminated terrestrial sites effectively tells us that small-rodent community HQ exercises are quite unnecessary anyway.) Where study outcomes are positive (i.e., repositioned rodents are observed to be unharmed from their imposed exposures), there can be appreciation for the experimental design that employed a universal feature of contaminant exposure studies, namely subjecting contaminant-free test subjects to chemicals. It is here that the study duration should be acclaimed, for it accounts for no less than 50% of the rodent life span, exceeding the exposure duration of a great many laboratory-based rodent-dosing efforts. The six-month exposure duration, in conjunction with afforded opportunities for contacting contaminants in natural settings and through natural behaviors only (incidental soil ingestion, preening, consumption of non-manipulated foodstuffs), facilitates the acquisition of toxicology information that is really needed. In truth, ERA is not interested in the shape of a chemical's dose-response curve, or in knowing the soil concentrations that correspond to acute and chronic effects. An ERA wants to know if the totality of a receptor's existence at a contaminated site, reflecting the varied chemical concentrations of multiple chemicals entrained in multiple media at multiple locations within a variable landscape, is health-impacting.

To the extent that transpositioned rodents confined to contaminated environs fare well through their testing, any lingering concerns are *de facto* limited to the tissue concentrations that the rodents may have assimilated. Presumably, this specific interest is directed at tracking the contaminated dietary exposures of carnivores situated at or near the sites, and potentially preying upon rodents with their newly assimilated tissue concentrations. The reader is reminded that an animal assimilating tissue concentrations of chemicals does not constitute an effect and, further, there presently exists no database that links blood, organ, or whole-body chemical concentrations to health effects (see Study 8). As for the prospect of contaminated rodents funneling their tissue assimilations onto higher food-chain elements, the reader should recognize that such contaminant transfers have, with certainty, been occurring for decades already at sites; the subject study, when implemented, then, will certainly not be first enabling the dietary contaminant exposures of local carnivores. Although the study design is potentially areally restrictive to rodents, the author's anticipation is that six-month tagged animals at contaminated sites will not bear any overt or internal (clinical) signs of illness.

The analytical work specified in the guidelines can inform on the extent to which chemical uptake occurs and the identification of those somatic compartments where chemicals pool. The tissue and whole-body measures of the additional animals trapped at the reference location(s) (guideline 6) will provide an important body burden baseline. The guidelines foster learning of an assimilation rate over a six-month period, and over multiple generations of a site's history ("c" of guideline 9). Guideline 8 provides an opportunity to observe any chemical depuration that might occur through repositioning animals from sites and confining them at a chemical-free location. The tabularizing of the tissue measures should lead to an enhanced understanding of the capacities of different chemicals, across an array of concentrations and amidst a variable context of contaminant mixtures, to assimilate and depurate. The larger the universe of sites submitting to this study, the more identifiable will be trends in chemical uptake (despite the uniqueness of individual site chemical footprints).

Study #11

Expand GPS Technology-Based Spatial Movements-Tracking for Mammals

Premise

If there is one overarching goal for the structure of ERAs, it is (or should be) to restrict their focus to just those species and their surrogates that are relevant to the site setting. Relevant in this study's context refers to a receptor having a sufficient physical presence such that it could reasonably receive enough of a toxic load to impinge on its health. Of course, a sufficiency of animals of a given type also needs to be present at the discrete contaminated sites that this book considers (and not just two individuals), and, for the purposes of the subject study, we will assume that enough species representatives do so populate these locations. The ample and freely available home range and related animal distribution information that exists today should be all that's needed to decide if a species or surrogate is spatially relevant in our context. Evidently, the available information, even as we have it accessible in regulatory guidance, isn't good enough, for we routinely find mammals with considerable home ranges being evaluated in ERAs for tiny sites. It's hard to know why this occurs. Seemingly, ecological risk assessors see the guidance-provided home range information as something to be woven into their assessments. Ignoring the fact that home ranges that vastly exceed the site size are open indications of insufficient site contact occurring, they interpret the supplied information as an inducement to compute area-use factors (AUFs). While the assessors' math supporting the AUFs might be correct (i.e., indeed, for example, a species with a 400-acre home range situated at a 4-acre site has an AUF of 0.01), we should find ourselves dumbfounded: why would anyone need to know of miniscule site-based HQs of wide-ranging animals?

This suggested study, unlike the others of this compendium, is an oddity. If ERA practitioners would properly apply the biological information we have at our disposal, there'd be no need to gather the data that this study encourages be accessed. (Note: the author, when presenting data that have already been generated with the study design to be described, has been fully forthright when referring to such work as "studies that didn't need to be done." The earlier forays into the study design were done with the purpose, so to say, of "hitting people over the head" to make clear to them what should be intuitively obvious, i.e., that there is no point to evaluating, say, a white-tailed deer at a 2-acre site, or a red fox at a 5-acre site.) Once again, it bears mention that there is no guidance-born requirement that at least one receptor of every kind/guild be evaluated within an ERA. How the unfortunate practice began of including in an ERA a small, medium-size, and large mammalian herbivore; a small, medium-size, and large mammalian carnivore; and a small, medium-size, and large mammalian omnivore, is unknown. Perhaps the intention was to be thorough, but the fact remains that ERAs have become unduly cluttered with animal reporting that doesn't at all inform. Taking a cautionary step back from the process can allow an ERA practitioner to appreciate that there are great gains to be had by simplifying ERAs through keeping receptor lists short and uncomplicated. This study encourages individuals to directly measure the time allocation of acknowledged wide-ranging mammals over their total areas utilized (TAUs). Such information will either support the incorporation of these animals at commonplace, relatively small terrestrial sites, or it will not.

Study Guidelines

1. Compile a list of wide-ranging mammals that, often enough, are conventionally assessed (i.e., with HQs) in ERAs for terrestrial sites. Note: although the white-tailed deer and the gray fox have already once submitted to the tenets of this study, interested parties are encouraged to select them again. Thus far, the white-tailed deer has been studied in two US states, and the gray fox in only one. Perchance, time-allocation statistics to be acquired may differ from what has been shown to this point, and newer information can usefully add to our knowledge base.

2. Establish that GPS-technology tracking devices (typically collars) are light enough in weight such that animals to be monitored are not encumbered in any way. Note: it may be that excessive tracking device weight arises only as an issue where the intent is to track for relatively long periods (perhaps of one year). A study, then, may become potentially workable if lesser-weight batteries are used to power the GPS units

for tracking periods of shorter duration (e.g., six months instead of an intended calendar year).

3. Suitable areas to trap and monitor animals are those where the habitat is unimpeded by human disturbances. Stagger monitoring periods, especially where only portions of the year will be evaluated, to ensure data collection through all months. Note: due to the nature of animal trapping efforts (e.g., less-than-ideal trapping success, malfunctioning tracking devices, etc.), monitoring periods will *de facto* be staggered.

4. Ensure that the spatial movements of males and females of a species are monitored.

5. Based on information provided in the tracking device's user's guide, program a sampling rate (i.e., the frequency with which latitude/longitude locations will be recorded) that will ensure battery life holding up throughout the intended monitoring period.

6. Plot all of an animal's recorded latitude/longitude locations using an appropriate (and often internet-available) GIS program. Use a suitable program to construct the minimum convex polygon (a.k.a. the TAU). Proceed to grid TAUs into uniform cells of the sizes of typical Superfund-type sites, with recommended grid cells being of 1, 5, 10, and 20 acres. For each animal, determine the percentage of its total recorded latitude/longitude locations that correspond to each cell at each scale of analysis. Recognize that the study's plausible assumption is that the percentage of animal's recorded locations that fall within a given cell is a reasonable approximation of the percentage of time that an animal allocates to that space. (Expect to find that significant portions of the TAU show little or no occupancy. Recall, though, that in the periods between locations being recorded [of perhaps 2 or 3 hours depending on the sampling rate], animals may have most definitely utilized/occupied what appear to be empty or hardly populated cells.)

7. For each animal, identify the three most utilized cells at each of the scales. Compute the percentages of the animal's time spent at each cell, and the average of the three-high cell percentages.

8. For each animal at the 10 and 20-acre scales, compute the probability that an animal's most utilized cell could (also) be a contaminated site. Do this by dividing the number 1 (representing the one contaminated site that could fall within the animal's TAU) by the number of cells formed in gridding the TAU. (Study enthusiasts should understand that contaminated sites do not frequently occur in a spatial plane, even where all regulatory programs that address contaminated sites are combined. Relative to the subject study, where the home ranges of the mammals to potentially serve as test subjects extend to multiple hundreds of acres, there could be, at most, but one contaminated site.)

Study Outcomes and Applications Thereof

The premise above spoke of the "oddity" of the subject study from the vantage point of readily available information and common sense, making it
intuitively obvious that wide-ranging mammals are not spatially relevant
for typical ERAs. In truth, therefore, no spatial movements-tracking study
is needed to begin with. For a great many of the studies described in this
compendium, we might have a good hunch as to what the outcomes will
be, but only a review of the data that the guidelines call for can confirm
that the thinking was correct. In contrast, the one-of-a-kind information
to be collected for the subject study cannot help but show what all should
suspect. Wide-ranging mammals cannot betray their biologically dictated
behaviors. No wide-ranging mammal, therefore, will spend an inordinate
amount of its time in a particular subarea of its home range or TAU. We
are hopefully not so naïve as to expect that an animal with a multiple-
hundred-acre home range or still-larger TAU will uniformly apportion its
time across such an expanse. That is, if a TAU is divided into 100 cells of
equal size, we should not expect the animal to have used each cell 1% of
the time. We should also not be so naïve as to expect dramatically skewed
time apportionment, as in a fox utilizing a 10-acre area 90% of its time,
and spending the remaining 10% of its time utilizing the other 390 acres
of its home range. Importantly, there are no supports for contentions of
the former; no information exists to say that animals are drawn to contamination, utilizing olfaction or some other sense to lead them to spend
disproportionate amounts of time in a chemically tainted milieu. For this
study we should expect to find large portions of a TAU hardly used, and a
handful of cells (even at the 20 or 30-acre scale) maximally used (although
"occupied" is a more realistic term) perhaps 15–20% of the time. With
just this much information gathered, it will be clear with actual field data
that the investigated mammals are not spatially relevant for study. Being
present at a site-sized area 20% of the time means that the animal does
not utilize that space (that some would prefer to claim could nevertheless
be contaminated) some 80% of the time. With such a demonstrated land-
use pattern, a critical receptor-of-concern selection criterion—having
a high degree of site fidelity—is surely not being met. Assuming study
researchers arrive at such area-use patterns with the species they monitor, they should be prepared for (great) resistance to hearing that the
animals are inappropriate for review within ERAs (on the grounds that
they are spatially irrelevant). Spending only 18% of its days at a contaminated site (particularly when a wide-ranging species does not ordinarily

feed continuously) does not make for a case of high or appreciable site fidelity (literally, site contact). Researchers, in an effort to gain traction with those who are adversarial, might want to put forward their findings through analogy, as follows. An intended animal-dosing study was to have Chemical X delivered for 90 days, a scheme that could fully be characterized as one of chronic exposure. As it turns out, either due to laboratory technician error or a short supply of Chemical X, dosing occurred over only 18 days, interspersed over a 3-month calendar. All should agree that the study could certainly not be termed chronic, and an honest principal investigator would not expect to find the anticipated effects that a 90-day exposure was said to elicit. We would trust, too, that the principal investigator would not try to portray this research as having achieved its aim.

Could it be, as a regulator might contend, that each animal's most occupied cell is a contaminated one? Study guideline 8 provides an answer to this question. Co-occurrence probabilities (of most utilized sites being contaminated ones) of 1 or 2% will handily prove that wide-ranging mammal inclusion is pointless. Not that any further cementing of the point is needed, but a consideration of the number of species representatives that could be expected to regularly inhabit a 20-acre parcel would further illustrate the senselessness of wide-ranging mammal consideration at typical sites.

Regardless of wide-ranging mammals being shown to be spatially relevant or not, the study brings forward a workable, direct, and site-specific methodology for defensible receptor-of-concern selection in ERAs for a distinct category of mammal. Appropriately showcased, the regulatory agencies should come to recognize that spatial movements-tracking can be a *bona fide* screening tool in this regard, particularly where it is recalled that an ERA goal is to streamline content. Updated ERA guidance could come to formally call for such screening. Where regulators can so acquiesce, opportunities to lobby for further ERA simplification in the form of excluding still other mammalian forms (for reasons other than spatial irrelevance; see Studies #26 and 27) can be discovered.

Study #12

Chemically Dose Terrestrial Environments in Support of a Longitudinal Health-Effects Study for Ecological Receptors

Premise

There are numerous evident disconnects between the chemical exposures that ecological receptors receive in the wild (at contaminated locations) and the ones that test animals receive in the laboratory. There can be no disagreement here, it is sincerely hoped. While there may be purpose to the animal-dosing studies of research ecotoxicologists—generally to develop TRVs and the like—the myriad ways in which the administered chemical dosing differs from the real-world chemical exposure case makes it that any utility at all within an ERA context is lost. In an ideal world there would be no disagreement on this, but the reality is that the point just made is outright anathema to the ears of the rank-and-file of ERA practitioners. We might envision a challenging and complex dialogue ensuing on the topic of the various aspects of laboratory testing in support of TRV development, and the ranking of these in terms of the degree to which each detracts from would-be usable toxicity values. Rather than provide a review (even a reduced one) of dosing study elements that stand as departures from actual chemical exposures, the aim of this study is to investigate the role of a single element of the mix—time. It is the short-term responses (e.g., 90 days) of test animals in the laboratory that ecotoxicologists study, but the ecological receptor in the wild reflects tens and in some cases hundreds of complete lifetimes lived within a milieu of chemical contamination that dates back to the start of a series of chemical releases. In the decades elapsing since a site became contaminated, many changes to the ecosystem have been occurring in tandem. At a minimum for a terrestrial ecosystem, chemicals have become sequestered and less bioavailable in the soil matrix, physically altered, and very possibly rendered less toxic. While these changes

were occurring, time allowed for receptors to develop chemical toler-
ances and other coping mechanisms, facilitated in part by the chemical
loadings accumulating in gradual fashion. The upshot of the foregoing
is that standard animal-dosing studies do not relate sought after toxicity
information, although ecological risk assessors often fail to recognize
this truth. Alternatively said, the standard animal-dosing study informs
only on how animals with no prior chemical exposure respond toxicologically when force-fed a singular chemical for a fixed number of con-
secutive days over just a portion of their lives. Glaringly absent and so
costly within a health evaluative context, animal testing to support TRV
development does not incorporate the phenomena of chemical weathering, altered chemical toxicity, and animal adaptation.

The present study sets forth the opportunity to create contaminated
sites that simulate ecological arrangements that have unfortunately come
about through poor environmental stewardship, and at which ecological
responses can be monitored over a protracted time course. Interested parties should understand that there will be constraints on the number and
sizes of sites that can be created. These parties are therefore encouraged to
collaborate with their peers to maximize the gains associated with har-
vesting the potentially voluminous and highly informing, one-of-a-kind
data to be generated.

Study Guidelines

1. Establishing the location where a contaminated terrestrial site can be
   created need not be an insurmountable or particularly challenging task.
   Two routes are available to researchers. First, they can approach established experimental forests, asking if their work can be accommodated/
   absorbed by the pre-existing facility. It could even be that what stands
   to be learned from chemically dosing an area meshes well with prior
   or ongoing research conducted at the facility. Researchers, alternatively,
   may find that they are free to deliberately apply contaminants to property
   they personally own. (To the extent that land-purchase costs are excessive, the study encourages organized groups [e.g., university depart-
   ments] to acquire the properties. Since this study, as with many others of
   this compendium, fosters partnering of enthusiastic scientists, it may be
   that land parcel-purchase costs can be raised where like-minded individuals combine their contributions or collectively fund-raise.) It goes
   without saying that the rights of the landowner as stated in the deed of
   sale should first be reviewed by licensed and competent attorneys before
   commencing the research effort. Clearly, the pivotal concern would be

one of "applied contaminant carry-over" to tangential properties and the populace accessing same. In this regard, one specific route of exposure in particular, contaminated groundwater use, must be given due consideration. That is, the concern over contaminants migrating away from dosed areas are not limited to what may become windborne or what may come to migrate via overland flow. As the reader surely recognizes, contaminants regularly percolate through the soil column and, often over time, come to diffuse into one or more aquifers. Once entrained in groundwater, the potential for contaminants to migrate considerable distances sets the stage for unintended but nevertheless potentially harmful human exposures occurring through water use (e.g., drinking, showering). It should be noted that just because contaminants will be applied to the soil, does not have to mean that contaminant sourcing to groundwater will occur. It is also possible that contaminant sourcing to groundwater is a non-issue in the region of interest due to the presence of low-yield aquifers that would not be able to support a community's water needs.

2. Enthused parties are reminded that typical contaminated sites (the focus of this book) are, relatively speaking, areally small. In that case, where perhaps 5–7 acres are to be contaminant-dosed, and the owned (private) property extends to perhaps 15 acres or more, it would seemingly make the best sense to dose a centralized area of the larger parcel. Such would minimize the chances of off-site migration. See guideline 5 for other site-size considerations.

3. A deliberate effort should be made to dose sites in gradual fashion, and certainly not through a singular application event. Researchers will recall that, when establishing study sites, the intent, wherever possible, is to duplicate the manner in which actual Superfund-type sites have formed, despite the reality that certain data are to be collected in the short-term. With the study to be a longitudinal one, and where, for example, data are to be collected at 5 or 10-year intervals, researchers should not be tempted to quickly establish the site.

4. To the extent that a created site should mimic a real Superfund-type site, several chemicals should be applied. Although the researcher is free to choose the contaminants and the pace at which they are released, etc., a suggestion is to endeavor to establish sites that highly parallel actual ones. (Complete histories of established National Priorities List [NPL] and other sites can be accessed to support such efforts.) Newly created sites will, of course, afford the unique opportunity to observe the earliest of ecological responses to introduced chemical stressors, should there be any. To the extent practicable, and independent of trying to recreate a formally listed site, contaminants should be released in a fashion reminiscent of the uncontrolled hazardous waste releases of Superfund and related sites. As a rule, researchers should become

fluent with site histories, and, where site re-creation is a goal, detailed descriptions of uncontrolled releases should be sought. By way of example, while simply sprinkling commercially available powdered silver onto the ground on multiple occasions will leave over a contaminant footprint, no silver-contaminated site has ever been so formed. Silver-contaminated (soil) sites though, have formed from the deliberate deposition of large volumes of (silver-containing) photographic laboratory processing ("bath") water onto the ground (as has occurred on federal properties, in designated areas not far from photographic processing facilities).

5. It is imperative that, prior to dosing, selected sites be rigorously characterized. This task will necessarily be supplying the baseline information that is virtually always missing in ERA investigations; understandably of course, deliberate polluters do not set about to characterize the areas at which they choose to contaminate. Task elements include:
   - standard measures of vegetative cover;
   - earthworm community characterization (worm weights, density, biomass/unit area, etc.);
   - compiling a thorough plant species list;
   - catch-and-release small rodent-trapping to characterize this animal group (counts, age distribution [juvenile, sub-adult, adult], sex ratio, body weight by sex and age);
   - bird community depiction (species list, bird and nest counts);
   - compiling a mammal species list;
   - estimates of the numbers of each mammal species that regularly uses the site.

   Importantly, this guideline will undoubtedly give interested parties pause to consider whether there is serviceable value to dosing sites of just 5 or 10 acres. Related to this notion, researchers should understand that at smaller sites observable/measurable trends in ecological response or adjustment to contaminated surroundings may only be possible for vegetation and small rodents (see next guideline). While such gathered data may be valuable in an academic sense, it may not supply anything to constructively assist the ERA knowledge base. Where interested parties take this guideline to heart, they may well find themselves seeking out larger study properties.

6. Although contaminated sites are virtually never remediated in order to afford protection to small rodents, there is purpose nevertheless in monitoring these animals longitudinally at a constructed (dosed) site. Small rodents are, firstly, fixtures of terrestrial ERAs. Because they are effectively tied to contaminated sites and thereby maximally exposed, they stand the best chance of having their health shortchanged. Even where study sites will be small, catch-and-release of rodents should proceed every year for the first five years after (site) dosing.

7. The monitoring data to be collected yearly for the first 5 or so years (during which time site dosing will likely still be occurring), and at subsequent 3–5-year intervals, should parallel that of the baseline dataset. Monitoring data should (obviously) not include somatic measures that necessitate euthanization, for the intent is to monitor populations where any shifts or directly observable physical changes are ecosystem-run only. Additionally, culling animals can likely interfere with the intended successful animal captures of subsequent monitoring events; in part, animals are vitally needed for young production. Researchers should recall that, while somatic measures may illustrate trends over time (e.g., livers are found to increase in size), critical second-order toxicology information is lacking for virtually all of these (i.e., it is not known how much of a change in a measure need occur before an animal is in a reduced health state).

8. Tabularize the data chronologically. Identify all instances of statistical shifts in routinely collected data (i.e., in body weights and population size for worms, birds, and mammals), and observable differences in external appearance. For shifts that are noted, review the longitudinal data for any evidence of recovery, i.e., where body weights, animal numbers, or health appearance return to what they were in the site's pre-dosing condition.

Study Outcomes and Applications Thereof

ERA concerns for contaminated sites that work their way through remedial programs are focused exclusively on receptor health. While some might wish that ERAs bore the sophistication to speak to the efficacy with which a site's "ecology" functions, ERAs do no more than feebly anticipate how certain site animals, as surrogates for others, fare. The expectation with ERAs is that sites should be able to support the local species, and in appropriate numbers. ERAs do not (and certainly should not) investigate somatic changes that are evident only upon necropsy, such as organ-to-body weight changes, shifted cytochrome P450 levels, or the presence of glomerular lesions and the like that only become known through histological review. If ERAs do explore such things, ERA practitioners are powerless to assign health rankings to animals that display them. Importantly, internal changes are not detectable by conspecifics and other site species, and do not correlate with animal behavior being any different from the norm, so far as we know. The more years a site has been contaminated, the more likely it is that hidden/internal somatic changes, such as those just mentioned, arose longer and longer ago. (It is hoped that ERA practitioners

are not so naïve as to think that, coincidentally, internal somatic differences first crop up in the very calendar year that field assessment work in
support of ERAs occurs.) Discovered somatic effects are distracting "red
herrings"; ERA practitioners are often wont to explore their underlying
mechanisms though they lack the evaluative thresholds-for-effect to do so,
and they are wont to recommend taking remedial action because of them,
although sound bases for such preferences cannot be expressed.

The subject study is structured to allow for observing if and when species numbers decline at contaminated sites. If there are declines, the study
is there also to chart the pace at which population replenishment occurs,
assuming that such a phenomenon is operative. Finding that site chemical
exposures impinge on biota necessitates investigating a window that ERAs
never consider, as follows. Terrestrial sites are always several decades old
by the time they submit to ERA. Alternatively said, terrestrial ERAs never
occur within the first year or even the first five years of a site becoming contaminated. It would be useful to know if, in the initial period pursuant to
a site becoming contaminated, animal and plant losses are sustained. The
subject study has the potential to uncover this extremely valuable information. If short-term population impacts are not sustained, at a minimum
we will be empowered to definitively quell an oft-expressed sentiment.
That is, ERA practitioners are wont to argue that remediation is necessary
at (typically decades-old) sites in order to preclude potential severe population impacts from happening. To the extent that the common finding for
this study's created chemically contaminated sites is that impacts do not
occur even in the initial post-contamination years, the expressed worry
of ERA practitioners will be seen to be unfounded. If population declines
do not occur when contamination is new, when opportunities to adjust
haven't yet been afforded the ecological receptors, the declines won't be
occurring 40, 50, or more years later either. Where the crafted sites illustrate that population impacts *do* occur in the short term and then peter
out perhaps a decade or two later, we also stand to learn that expressed
concerns over population impacts at conventional sites are unfounded. At
a minimum, such information will help to resolve an element of the theoretical underpinnings of RSA, ERA's only direct health status assessment
method. All RSA applications to date have demonstrated that mammalian reproduction is not corrupted at contaminated sites. Although discussion on past/historical site risks is a moot topic in ERA, site-specific
interpretations of RSA outcomes still to this time leave open the possibility that reproductive impacts might have flourished at sites decades ago.
Potentially, the information to be brought forward, particularly where a
multitude of artificially created sites are studied, will inform that even in

the early years of a Superfund site (i.e., even before site discovery), reproductive impacts never took hold.

As alluded to in guideline 5, the study will put researchers and those digesting the study's findings in touch with the spatial requirements of sites submitting to ERA. Presently, droves of ERA practitioners are fully inattentive to the spatial dynamics of sites, and examples (provided below) abound. An anticipated pay-off of the study is its capacity to cause practitioners to come to grips with our errant HQ-based ERA focus, and our general inabilities to produce meaningful assessments.

The study should trigger the following considerations:

1. Many will insist that ERAs proceed for 2-acre sites, but they cannot articulate what could go wrong ecologically at sites so small.
2. Few (state agencies and others) opine on a minimum site size at which the ecology could be damaged, and where it would be prudent to intervene.
3. Few, if any, are able to provide an explanation for a 350-acre property becoming contaminated through-and-through.
4. Small rodents as sentinels never seem to be impacted at contaminated sites, implying that other (larger) mammals at sites are undoubtedly free of impacts as well. Since (a) HQs are so obviously erroneous, particularly for the mammals that we are able to regularly collect, and (b) we cannot chance HQ estimation to inform on larger forms, what are we to do?
5. How large would a site need to be to furnish a sufficiency of birds or mammals (other than rodents) to make worthy their direct assessment for overall health and population stability?

Study #13

Validate Biokinetic Uptake Modeling for Freshwater Fish

Premise

Biokinetic models for aquatic environments seek to estimate assimilated chemical concentrations in food chain organisms, an example being those of a 3-step food chain consisting of periphyton, mayflies, and minnows. Model estimations are a function of the chemical concentrations in the water and diet to which an organism is exposed, with consideration given as well to the time-course over which chemical uptake and elimination processes occur. The primary focus with biokinetic modeling in an ERA context is, of course, fish, as tainted fish have the potential to pose threats to still-higher chain piscivorous forms such as osprey and mink. Hence, for the protection of the lower-chain species themselves (such as the minnows in the above example), there is no need to know if model predictions are on- or off-target. As has already been mentioned in this compendium, information is lacking on the magnitude of body burdens (as in whole-body measures) that signify health impacts to the fish (or other species, for that matter) that bear them—an unfortunate arrangement. Nevertheless, we have a need to know of the whole-body concentrations of larger fish for the potential threats that follow from their serving as prey. As with all environmental models, there is great potential to generate widely inaccurate results. Gross under- or overestimates can follow from model oversimplification and an inherent failure to understand the nature of chemical transformations, with these likely paling in comparison to the actions of stochastic factors that lie beyond the modeler's control.

Opportunities to evaluate the accuracy of biokinetic model prediction in larger fish surely exist, and the subject study is attentive to this. In the main, this study is asking enthused parties to document whole-body chemical concentrations in larger fish as they exist in waterbodies that have received anthropogenic inputs and that have earned places on lists of sites to submit to ERA.

Study Guidelines

1. No part of the study that entails the exposure of fish to contaminated surroundings is to occur in a laboratory or other unnatural setting.
2. Compile a list of contaminated freshwater sites that have yet to be remediated in some fashion (e.g., dewatering, sediment removal). For each site, chart the chemical(s) that pose purported ecological issues and the range and mean concentrations for each in the water and in the sediment's bioactive zone. Acknowledge that no two sites are likely to be the same in terms of their contaminant holdings; chemical suites and concentration ranges will notably vary.
3. For each site, compile a list of the largest fish species present (e.g., bluegill, trout) and specifically those that would be consumed by the larger piscivores that garner concerns that adequate protection from site-specific chemical exposures might not be afforded (e.g., mink). Collect a representative sampling of fish and record fish weights and lengths. (Sites with fish no larger than minnows cannot serve as viable sites; osprey and mink and other piscivores that rank high with critical protection needs do not principally feed on minnows and the like.)
4. Collect site-specific food-chain information (e.g., measured tissue concentrations of the diet items of the larger fish under study), and apply appropriate biokinetic uptake models (based on literature-based chemical transfer rates and such) to estimate the whole-body concentrations of site chemicals in the site-specific fish species that will be part of the study.
5. For each viable contaminated site, identify one or more non-contaminated freshwater locations that can serve as suitable references. Endeavor to secure paired reference locations that support the same larger fish species present at the sites. Document those instances where otherwise-viable reference locations support species that are (only) similar to the suitable larger fish species of interest at the sites.
6. After securing IACUC approvals, collect from each of the reference locations a minimum of ten specimens for as many species of larger fish (e.g., bluegill) as is workable, and run whole-body analytics on these. (This will ensure that baseline body burdens are understood, thereby supporting the guidelines that follow.)
7. Collect an additional 20 fish of as many appropriate larger species as a reference site will supply, i.e., those defined in the preceding guideline. Tag these fish and transfer them to their paired contaminated site

waters. After six months, use electro-shock or other means to catch half of the site-introduced fish (i.e., ten fish) of each species, and have whole-body analytics run on them. At one year post-introduction to the site waters, collect the remaining (ten) introduced fish and have these analyzed for whole-body concentrations as well. For all retrieved fish, assess overall health/appearance, and, at necropsy, prior to whole-body sample preparation, record any observed internal physiological anomalies.

8. For each of a site's established contaminants (guideline 2), document increases in the whole-body concentrations of the post-introduced fish relative to the baseline whole-body measure (guideline 6).

9. Compare biokinetic model-estimated (whole-body) concentrations of site-specific chemicals to those directly measured in the post-introduced fish. Tabulate all statistically significant larger fish species' whole-body concentration differences with site-specific chemical concentrations in water and sediment.

10. Compile a list of waterbodies (e.g., lakes) with catch-and-release-only fishing programs, because the fish at these are said to be unhealthy for consumption due to their chemical body burdens. Gather the chemistry and other information to support biokinetic uptake-modeling to the larger resident fish forms (e.g., water and sediment contaminant concentrations, identified food chain elements).

11. For waterbodies that already have a stocking program in place, determine whole-body concentrations of the fish to be stocked, and tag the fish prior to release. Collect tagged fish at six-month intervals; following a general health condition review, run whole-body analytics on representative samples (minimum 10) and compare the recorded concentrations to both the pre-stocking ones and to all the half-year whole-body estimates. Identify all instances where, for matching time points, the modeled concentrations approach or exceed the actual measures. Document trends in the whole-body concentrations (seemingly, increases in these) over time.

12. At viable sites, such as are described in guideline 2, endeavor to implement a fish stocking program, and apply guidelines 8 through 11.

Study Outcomes and Applications Thereof

Direct analytical measure provides the truest assessment of a fish's body store of assimilated chemicals. It follows that in any and all ERA work involving chemical fish uptake concerns, we can never be fully reliant on biokinetic modeling. The truest understandings of whole-body chemical assimilation rates and the magnitudes of concentrations achieved can only come from natural as opposed to laboratory settings, and where fish without burdens are introduced into the former. Preceding the present discussion, it does not appear that anyone working in ERA or the related ecotoxicology field has suggested, let alone actually pursued, introducing contaminant-free fish to contaminated waterbodies as a means of securing truthful bioaccumulation information. (Fish stocking programs to allow for recreational fishing, be it catch-and-release or otherwise, does not approach what's being discussed here.) The procedure though, is absolutely essential. Releasing chemically free trout to a contaminated lake, for example, stands to inform on the phenomenon of chemical biouptake to a vastly greater degree than does exposing laboratory-reared mosquitofish to five-gallon aquaria with amended concentrations of singular chemicals, for periods of just two or three weeks.

This study should, at a minimum, either bear out the utility of biokinetic modeling with fish, or indicate limitations to the approach and how severe these run. Potentially, biokinetic modeling "works" with simple(r) aquatic systems, i.e., those with only one or perhaps two chemicals present. The study design involving actual contaminated waterbodies is, in part, designed to bring this point to the fore. There is little need for models that can only perform well with contaminant suites as brief as these, given that, typically, multiple anthropogenically released chemicals define a site's contaminated water column and sediment. Models that can predict biouptake fairly well for one chemical, but predict poorly for a second or a third, are really worthless; we would always have to be second-guessing the predicted figures. This leads to an overarching observation. With the potential for specious model output, it should be clear that direct analytical measure is always needed. What advantages, then, are there to biokinetic modeling, and why should scientists continue to work at improving its lot? Alternatively said, how hard is it to catch some fish and thereby get at the truth with any scant uncertainty traceable only to the number of specimens analyzed? Perhaps the (only) answer concerns the case where projections are needed for the out-years. One example of this would be the need to know if continued exposure of fish to non-remediated waters will, in time, lead to their assimilating concentrations that threaten their consumers. A second example would pertain to permitted (Resource Conservation and Recovery Act) facilities, and the need to know of a safe long-term discharge rate of a contaminant-bearing effluent to a receiving waterway.

Realistically, it cannot be that ERA practitioners might harbor concerns that assimilated concentrations are unbounded. The literature is replete with references illustrating that equilibrium between a contaminant's environmental concentration and that of the whole-body fish is established in a relatively brief time frame, perhaps just over a matter of weeks. Although an iffy surrogate, hexane-filled dialysis bags suspended in organics-contaminated waterbodies have demonstrated same many times over. For waterbodies that have been contaminated for years then, as is the common case and the one to which this book is directed, we have even more reason to stay with direct analytical work as the method of choice, and to see the lesser need for biokinetic modeling of whole-body concentrations in larger fish. The argument that follows parallels that expressed in a number of other study descriptions of this compendium. Whatever maximum assimilated concentrations in fish should be, these have been achieved by many a fish in a given waterbody over a span of decades. (We recall that contaminated aquatic sites, no different from terrestrial ones, first submit to ERA work decades after having been contaminated.) With this reality, piscivores (e.g., gar, merganser, mink) have then, too, been ingesting unhealthful contaminant loads for an equally long time. If the piscivores over which we are so concerned were to succumb through dietary fish intake, such would have occurred already. Where whole-body contaminant loads linger over considerable periods of time, ERA practitioners need to be answerable in two ways. They must be able to explain lacking accounts of fallen piscivore-of-concern numbers at sites and their immediate environs. They must also be able to explain the presence of readily observable piscivores at sites in the present day. In a sense much larger than validating the capacity of biokinetic models for fish, this study is calling for a re-assessment of the perceived need for certain aquatic ERA applications altogether.

Study #14

Adequately Field-Validate the Efficacy (Predictive Capability) of the Simultaneously Extracted Metals/Acid Volatile Sulfides (SEM/AVS) Method

Premise

The SEM/AVS method for predicting toxicological impacts to benthic organisms joined the library of sediment quality assessment techniques several decades ago. With theoretical underpinnings that are reasonable and straightforward, ERA practitioners had no problems latching on. Should they, though, have been so quick to adopt the technique that reputedly anticipates heavy metals in sediments causing toxicity to sediment-dwelling organisms? It is certainly true that heavy metals need to be in a bioavailable form to pose toxicity, and that metals in the solid metal-sulfide form (i.e., bound to sulfides) are rendered non-bioavailable. As opposed to metals that precipitate as a solid or that are sorbed to a sediment particle, those (metals) that are dissolved in the pore water are understood to be freely bioavailable and thereby most threatening in a toxicological sense. How an excess of bioavailable metal ions can come to take up presence in the pore water follows simply from chemical interactions occurring between (simultaneously extracted) metals and acid-volatile sulfides. The situation will arise when metal ions are present in quantities too great for sulfide ions to bind with them all (despite the latter's high affinity for heavy metals); hence, the subject method's focus on the ratio of the quantities of the two ion types (i.e., SEM relative to AVS).

For all of SEM/AVS's theoretical background and fine measurements of stoichiometric ratios, the method might not be an accurate toxicity predictor or even a predictor at all for the situation in which it is thought to be needed (i.e., out in the field). An unbiased review of the method's test trials leaves much to be desired. Newer students of sediment toxicology, who know of the method's workings but not of the method's history, might marvel at the paucity and type of studies that, back in the day, served to convince the ERA community that method application was purposeful. A well-referenced study from 1996 embellishes this discussion. It involved metal-spiked laboratory sediments where variable concentrations of AVS had already been established. The toxicological responses of benthic organisms released to the mix completed the picture. The results could be described as striking, and fully supportive of the basic theoretical SEM/AVS framework. Where the ratio was less than or equal to 1.0, only 1.1% of trials showed greater toxicity than did controls, and 73.5% of trials showed greater toxicity than controls where the ratio was above 1.0. It should be clear to the reader where this highlighted study leaves us wanting. Chances are that artificially spiked sediments in laboratory aquaria are not reflective of the contaminated sediments of natural settings. In line with the premises of a number of other studies in this book, in ERA we are not interested in the responses of test organisms that (rather suddenly) find themselves artificially placed into contaminated surroundings. Contaminated waterbodies that submit to ERA are decades old when studied, and there are no benthic forms arriving at contaminated sediments to experience contamination cold-turkey. It is rather the case that the benthic forms to first experience contamination are those that are born to benthic forms that have been living with contamination over generations and years. Perhaps it is only for artificially spiked sediments with pre-arranged AVS concentrations along with supplied commercially reared test organisms that the SEM/AVS framework holds up, but our interests are with tainted waterbodies in the out-of-doors. A review of so-called SEM/AVS "field studies" does not improve our lot for prospects of confidently knowing that the SEM/AVS ratio is a *bona fide* toxicity predictor. One commonly referenced study, also from 1996, evaluated *laboratory* exposures to field-contaminated sites, hence the use of "so-called" as a field study descriptor here. Interestingly, while the method was touted as a good toxicity predictor, a re-analysis found quite otherwise. Despite the foregoing accounts of the method's insecure foundations, we continue to apply SEM/AVS in ERA investigations.

With this study, the enlisted researcher is called upon to conduct the necessary work to validate the predictive ability of the SEM/AVS method at contaminated waterbodies. One truly legitimate reason to take on the work bears on the potential for ERA practitioners who apply the method to err in their interpretation of SEM/AVS outcomes. They can lose sight of the fact that a discovered ratio above 1.0 is merely a screening tool and nothing more. A ratio above 1.0 does not automatically mean that benthic forms have been wiped out. With the method as it presently stands, method predictions of unacceptable toxicity should, at best, serve only as triggers of site-specific field investigations.

Study Guidelines

1. Assemble a list of contaminated aquatic sites, both freshwater and salt water, where no sediment remediation work has occurred. Sites that have already had SEM/AVS run in relatively recent years, and where ratios exceeded 1.0, are most eligible.
2. Compute SEM/AVS ratios for those sites that, as yet, do not have them.
3. Develop a list of benthic community measurements to be evaluated, providing for each a defensible basis for its inclusion. Note that a shifted benthic community assemblage is not necessarily indicative of outright toxicity, as other factors (some perhaps physical in nature) may account for such a finding.
4. For all cases of SEM/AVS ratios above 1.0, conduct the benthic assessment work necessary to either confirm the method's predictive ability (namely that the benthic organisms are experiencing toxic effects) or conclude that benthic impacts are minimal or absent.
5. Ensure that the benthic assessment work is defensible. It should be clear that, other than counting and identifying species, no work is to take place in the laboratory, and no commercially reared test or other organisms are to be used.
6. For those instances where toxic effects to benthic species coincide with ratios above 1.0, indicate the extent of compromised waterbody ecology, and explain why the field-verified benthic toxicity observed matters in an ERA context. For this latter task, endeavor to ascertain approximately how longstanding is the manifested toxicological *in situ* condition.
7. Compute the percentage of the reviewed waterbodies for which the SEM/AVS method was an accurate predictor. Compute the percentage of reviewed waterbodies that had observable ecological impacts of any degree, and, especially, where these occurred at higher trophic levels. (Examples of impacts to track would include distinctly smaller fish populations, noticeably unhealthy fish, a low dissolved oxygen condition, or evident eutrophication related to the presence of chemicals in the sediment.)

Study Outcomes and Applications Thereof

Twenty or more years after the availability of a sanctioned predictive toxicity method is an odd time to be seeking a method's true validation. Field validation is nevertheless needed in this instance because the heretofore-supporting information for SEM/AVS is open to certain challenge. Is the survival of polychaetes and amphipods that are placed into glass canning jars for 10 days something we need to know about, and are the response data of such experimental arrangements reflective of benthic organism responses that occur in the contaminated sediments of the sites we investigate? Investing the energies to gather the data described above, regardless of what it should show, would likely stimulate thinking amongst the interested parties taking on the project, and of ERA practitioners reading about the findings as well (for the study presents a fair opportunity to publish the findings in the peer-reviewed literature). The potential for triggered stimulated thinking is perhaps the greatest gain for this study. Primarily to consider is that of the value added of attempting to predict if benthic forms are toxicologically offset where sediments have been contaminated for decades: "Why is benthic toxicity being investigated so many years after the fact?" and "Does it matter if benthic toxicity is demonstrated after so very long?" The study effort stands to stimulate broader thinking: "What purpose is served in testing any ecosystem, terrestrial or aquatic, when (a) so much time has lapsed since environmental media became contaminated, and (b) when any chemical-posed biological differences that might exist, can only be known if and when obscure samples are brought into the laboratory for examination (e.g., for organism counting and measurement recording)?" Presumably a laboratory testing-derived SEM/AVS ratio greater than 1.0 for collected sediments connotes the inability of site polychaetes, amphipods, and other species to survive in contaminated waterbodies, and further that benthic populations ceased to exist many years ago. With the described study, we stand to learn if such connotations bear out.

Presumably, benthic organisms are targeted with SEM/AVS for as many as two reasons, including these very organisms being important in that they allow for the proper functioning and sustainment of their ecosystems, and these organisms serving as surrogates for other aquatic species. The field validation study holds the potential to confirm or refute the reasons for benthic organism testing and evaluation. Beyond verifying the predictive abilities of SEM/AVS, the computations described in study guideline 7 bear the potential to either illustrate a continuing utility in assessing sediments in this fashion, or a basis for dispensing with method application. As with other sediment-based studies in this book, the collected data from this effort should contribute to the still-broader discussion on the merits of sediment review in ERA work altogether. ERA practitioners cannot truly expect to encounter fully biologically dead sediments at conventional contaminated sites. While the species assemblages might have shifted toward more pollution-tolerant forms, the benthos might not be ecologically shortchanged in the least, and the column water environment need not present as stressed. Stimulated thought on the merit of endeavoring to "repair" sediments when these afford the requisite biomass and other key ecosystem features they should, and despite species shifts having occurred, can lead to a realigned focus on the purpose of sediment investigation in ERA altogether.

Section II

Laboratory Studies

Drawing by Elana Barron.

Study #15

Establish a Database Supporting a "Top-Down" Medical Screening Scheme for Birds and Mammals in the Field

Premise

Ecological receptors at contaminated sites are routinely evaluated on the desktop with HQs and other simplistic benchmark comparisons. A distinctive feature of these so-called health evaluations and/or health effect projections is the absence of any direct (i.e., hands-on) contact with living forms, and particularly ones that concern us. This deliberate bottom-up approach to assessment leaves much to be desired, particularly when the highly imprecise HQ construct almost always paints a receptor as ingesting more than the safe level of a chemical. The implication of the desktop work ordinarily is that receptors are at risk of developing health effects, if they are not already displaying them. There are many reasons to think that ecological receptors at contaminated sites are not in the throes of succumbing to illness, however. For one thing, receptors have had decades of opportunities to adjust in various ways (e.g., physiologically, genetically, evolutionarily, etc.) to a contaminated site condition. It is also true that documented instances of health-compromised animals at contaminated sites are virtually unknown. Given this brief summary, it is reasonable to suggest that if site receptors should be health-compromised, evidence of the same should be observable and detectable upon direct physical examination or upon the cautious review of drawn metabolic samples. ERA practitioners may be disinclined to collect receptors from the wild for the purpose of assessing health in a manner akin to the way a doctor assesses a patient's health (i.e., from information gleaned in an office examination and from laboratory specimen analysis pursuant to an office visit), but trapping animals is easily accomplished. This study is an open invitation to enthused parties to develop a "top-down" medical screening scheme for birds and mammals that incur contaminant exposures; animals that would otherwise, at best, only have HQs computed for them.

In the main, the study calls for the development of a comprehensive database of the normal ranges of metabolic parameters that can be measured in the blood and urine of mammals and birds, paralleling what exists for human medicine. By way of example, an equivalent of human medicine's "basic metabolic panel" (BMP), which informs on kidney function, blood acid/base balance, and blood sugar level, can be reviewed in species like raccoon, hispid cotton rat, marsh wren, and pheasant. A routine BMP evaluates a standard set of some seven parameters, namely BUN (blood urea nitrogen), CO_2, creatinine, glucose, serum chloride, serum potassium, and serum sodium. It is understood that some of these may not occur in the blood of non-human species or that levels in animals are so low as to impede reliable measure or quantification. This study, it should be noted, is not designed to perfectly replicate the suite of reviewed parameters upon which physicians focus; it is rather designed to replicate the means by which physicians arrive at their health assessments (i.e., diagnoses of disease). Importantly, it is well within the realm of possibility to arrange for blood and (with an admittedly greater effort) urine to be collected from animals in the field. Securing IACUC approval should proceed without challenge, for the basis of the field data collections – to gather the information to support determinations of animal health at contaminated sites – is noble and sound. Additionally, numerous environmental investigations that involve minimally invasive biological sample collection, such as blood draws, are the order of the day in studies and other initiatives of natural resource trustees, and in ongoing scientific research overall.

Study Guidelines

1. Only new data is to be sought in establishing the database. Information reported in the available peer-reviewed literature cannot be workable, for it is unlikely to be sufficiently comprehensive in nature. By way of example, it is doubtful that, for any given measured parameter, the data of a representative number of animals is what is currently reported in encyclopedias.

2. The envisioned profile of metabolic and/or physiological measures to be assembled need not be limited to those that are assayed in human biological fluids. Through trial and error, a honed ("certified") list of parameters to be collected and entered into the database may not be

as inclusive as the list of routine and occasional parameters tested for in human samples. It may also be that several parameters that are not measured in humans come to populate the list. Scientists with specialties in comparative animal physiology, veterinary medicine, and hematology should be sought out to craft species-specific lists that should be the most utilitarian. Lists should not simply be made up of all those parameters for which a capability to accurately measure already exists. (For blood samples, it would seem prudent to have the database include: complete, red blood cell, white blood cell, and platelet counts; hemoglobin, hematocrit, mean corpuscular volume; calcium and electrolyte levels, and, to the extent practicable, an array of hormone and enzyme levels.) Importantly, the purpose in collecting parameters should always be defensible. Study researchers should be aware that when "certified" database measures are to be put to use (see Study #7), some or possibly many of the measures may turn out to not be particularly utilitarian.

3. Initially for a given animal species, the database should necessarily segregate parameter information by state or geographic district. Where a later detailed statistical review of the grouped data illustrates notable similarities (e.g., by region), it may be possible to combine the data of two or more states. Plausibly, the data for certain (or all) parameters can be pooled if dataset variability with regard to a given factor is found to be minimal.

4. Data need to be collected for males and females of each species. As with the previous guideline, the possibility may always exist for pooling the data of the discrete sets. A statistically defensible and achievable number of specimens should constitute each dataset. (Note: it is recognized *a priori* that, for any of the measures, knowledge of a parameter's natural variability, which would so greatly assist with determining appropriate sample sizes, will almost certainly not be known. Consider that, presently, the natural variability for any liver enzyme in the female cottontail is unknown.) See guideline 7.

5. Every effort should be made to standardize the acquisition of data. Using blood as an example, for a given species (e.g., skunk, raccoon), the collection of samples should standardize: the puncture site (e.g., a vein in the forearm), the type of collection tube (as heparinized or EDTA-containing), the blood volume collected, and the laboratory conducting the analyses. (Ideally, the same laboratory should conduct all of the analyses supporting the database, but this is highly unrealistic.) To the extent that samples can be held for longer periods because preservatives are used, and where, of course, measures are not compromised with extended sample holding, the possibility exists of having a series of samples (e.g., bloods for the gray fox [*Urocyon cinereoargenteus*] in the state of Georgia) all run at the same time under the same conditions.

6. The health-screening scheme to be developed need not only be biological fluid-based. Consider that humans are health-screened in a number of other ways (e.g., chest x-ray, blood pressure reading). While certain desirable tests and measures may not be easily implemented for animals, or may not be realistic at all, radiography should be a consideration for inclusion in the health screening scheme. Relatively transportable equipment to allow for x-rays is available to support such an endeavor. It is recognized that IACUC recommendations, and cost and other considerations, may not allow for the collection of radiographies and other less standard datatypes.

7. To the extent feasible, consideration should be given to compiling age-specific parameter measures, for physiological profiles might vary substantially with animal development or senescence. Importantly, and depending on the species, it may be that data for no more than two life stages (e.g., juvenile and adult) would ever be needed.

8. As should be apparent, this study presents a great opportunity for research teams to work in a collaborative atmosphere. Where scientific community research interests and granting institutions can support this endeavor, database assembly can be notably hastened.

Study Outcomes and Applications Thereof

Pursuant to a completed QA/QC effort, the assembled database will facilitate an intended new order of health assessment for the terrestrial ecological receptor (described in the next study). Additionally, the collected data will very likely provide insights into the commonalities or distinct differences in metabolism across sexes and species. In turn, this information may shed light on the design of other animal health initiatives (e.g., designing better diets for animals used in research, improving captive breeding program success rates, etc.). En route to achieving the ultimate goal of having in place, a health assessment scheme for animals in the field at contaminated sites, the study cannot help but compile an impressive library of animal biology information that can fuel research in an array of environmental and evolutionary biology sub-disciplines.

Study #16

Develop a Serviceable Effects Residue Database for Plants

Premise

It is a fair statement to say that the available means for assessing potential toxic effects to plants at contaminated sites are inadequate and underdeveloped. Should site vegetation be assessed at all within an ERA effort, this will take the form only of comparing contaminant concentrations in soil with tabular (i.e., "look-up") plant protection values, where exceedances of these are taken to mean that vegetation is stressed and unhealthy. One of the biggest problems with this approach is that contaminated sites, irrespective of the suite of chemicals their soils hold, are commonly vegetated where they should be, with bare spots notably infrequent and miniscule in size. With no more information than this, it can be argued that the need for vegetative assessment in ERA is probably not demonstrated altogether. That the most commonly applied compendium of plant protection benchmarks bears the caveat that the benchmarks are poor indicators of toxic effects where sites are adequately vegetated is unfortunate. The situation begs the question of why can't a plant-protection benchmark compendium be assembled where benchmark exceedances *do in fact* well predict the observed condition in the field (i.e., that plants are/will be found to be sickly and all but dead)? Why the benchmarks are so weak and non-applicable is easy enough to understand; they are often based on laboratory exposures that highly deviate from the site condition. The benchmarks are frequently for agricultural species, most unlikely to be found at sites (e.g., bush beans, cabbage), with many of the exposures of seeds placed into solution as opposed to soil, and for particularly brief periods, perhaps just seven days. Clearly, extrapolating from such testing arrangements cannot inform. The overarching question left outstanding asks of the purpose in developing plant-protection benchmarks when a site can be walked to ascertain if a site has vegetative cover and to document how extensive it is. A second lingering question asks if there can really be an expectation of impacted vegetation occurring at some point in the future, when impacts haven't arisen after decades. (Critically relevant to this discussion is a cautionary note to ERA practitioners. A partially or significantly different plant assemblage at a contaminated area relative to a reference location need not mean that soil contaminants are at play. It is more likely that historical site activities physically destroyed the vegetation at the former, leaving the site poised to be vegetated anew by plant species that easily establish themselves as pioneers.)

Vegetation impacts at contaminated sites can be of two kinds. Plants can either be precluded from rooting and populating the landscape (more correctly, the *sitescape*), or they can become compromised in the areas of growth and hardiness, the latter mediated by chemical uptake followed by chemical distribution to above-ground plant parts. With justification lacking for investigating impacts of the first kind (because cases of precluded vegetation at sites virtually never occur), it would seem prudent to establish plant residue levels that speak to compromised plant health. This study is intended to generate an answer to the question: Does it matter *to plants* that they harbor chemical residues that trace to the contaminated soils in which they stand? As with other described studies in this book, this one invites multiple interests to participate. Realistically, given the scope of this study (in terms of variable plant species and numerous chemicals), only through the combined contributions of many can a comprehensive answer to the essential question materialize.

Study Guidelines: Laboratory (Greenhouse or Plant Nursery Preferred)

1. The study guidelines are directed at manipulating conditions such that plants assimilate considerable chemical concentrations, and noting whether the artificially adjusted soil concentrations that result in plant residues actually occur *in nature* (i.e., in Superfund-type site soils). It is important to recognize that the study's interest is only in artificially inducing chemical residues, albeit exaggerated ones, where such occur through otherwise normal uptake and assimilation mechanisms. Thus, direct chemical application to the plant exterior, or chemical injection to any portion of a plant, must be excluded. While plant chemical uptake via aerial deposition is a legitimate pathway, it is likely that such does not lead to appreciable internalized concentrations (i.e., tissue residues).

Further, while good reasons exist for necessarily considering the ecological consequences of aerially deposited chemicals to plants (e.g., for the evaluation of ingestion pathway hazard to herbivores), the subject is directed only at assessing plants for their own sake.

2. Access ERAs from different U.S. regions and from different habitats within each, seeking out their terrestrial plant species lists. Assemble a compilation of herbaceous and woody species to research, all with the common feature of never having been used in toxicity testing of any type, thereby guaranteeing that they have never formed the basis of any plant-protection benchmarks.

3. Through extensive literature review and deductive reasoning, determine (reasonably) how long it might take for substantive chemical concentrations to appear in the stems and leaves for a broad matrix of chemical and plant pairings (e.g., arsenic and multiflora rose [*Rosa multiflora*]; chrysene and common spicebush [*Lindera benzoin*]).

4. In sizeable bins or test plots, arrange to culture and maintain an array of considerable-size or full-size (mature) plants, with multiple replicates of each. Vis-à-vis guidelines 5–7, some replicates are to serve as study controls.

5. Investigating single chemical exposures only, apply chemicals to the soil at a pace where it is hoped that root uptake followed by chemical transfer within the plant will occur. It should be clear that the soil dosing cannot occur at one time. Such would unfairly overload contaminant processing defenses, and also depart substantially from the real-world case; chemical releases that create contaminated sites do not occur on just one day. To minimize the number of replicates needed, adopt a range-finding approach to the soil-dosing that is to produce significant plant tissue residues.

6. After estimated lag times (guideline 3), periodically sample leaves of dosed-soil and control-soil plants for evidence that chemical uptake, preferably of the substantial degree sought, was achieved in the former. Where plants will not succumb to illness because of it, remove (occasional) smaller stems for testing as well.

7. Maintain a database of plant/chemical associations, study-established chemical soil concentration, manifested leaf/stem chemical concentration in plants of treated soils relative to soils of control plants, and the latency from soil application until achievement of above-ground plant chemical assimilation.

8. Monitor the overall health condition (height, coloration, turgidity, etc.) in all plants that manifested noteworthy tissue residues. Where resources permit, extend the plant health monitoring to the assessment of gas exchange efficiency and photosynthesis efficiency, or still other biochemically mediated essential biological functions.

Study Outcomes and Applications Thereof

At a minimum, the study's assembled information will definitively resolve the matter of the ecological relevance of manifested chemicals in above-ground plants parts. (The reader should appreciate that ERA does not concern itself with chemical-in-root contaminant concentrations, because ecological receptors of interest do not contact plant roots. Further, other than through the actions of feral hogs, perhaps, the roots of bushes, shrubs, and larger woody forms are never exposed such that animals could feed on these, in any case.) The study is designed to identify linkages, should there be any, of lesser plant health coinciding with areas of soil contamination. Where overt impacts are absent, there would, strictly speaking, be no need for reviewing plant tissue data. Where notably stressed plants are observed, the tissue data may inform on causation. Again, should it be that ecological receptors are, all the while, consuming plant parts (e.g., leaves) to potentially develop toxicological insults from this behavior, such is beyond the subject study's concern.

The study has the potential to resolve the question of whether or not ERAs need to involve themselves with plant (health) assessments altogether. The fact that unexplained bare soil patches occur so very infrequently at contaminated sites suggests that grasses and understory vegetation are all highly tolerant of contaminated soils, whether these plants uptake chemicals or not. The study will inform on how plants encountered in the field, and that otherwise make for the most uncommon of toxicity test species, fare in the face of imposed soil contamination, something that the enthused ERA practitioner should want to know. The results of the study may serve to reinforce the concept that the presence of contamination in plant tissue does not have to signify stress or impact. Such recalls the library of information, already dating back several decades, on plants as hyperaccumulators. That a number of plants belonging to distantly related families share the curious ability to grow on metalliferous soils and to accumulate extraordinarily high amounts of heavy metals without suffering phytotoxic effects suggests that still other plants can do something similar. (Importantly, where metal hyperaccumulating plants are removed from sitescapes after having achieved what was asked of them, such is only done to minimize potential human, and not animal, exposures. Hyperaccumulators do not appear any worse the wear for their soil metal-extraction efforts.) Still with regard to metals, we are aware that more than 500 species of flowering plants have been identified as having the ability to hyperaccumulate. While common basement-layer and understory plants might not be categorized as hyperaccumulators of the kind applied in phytoremediation efforts (such as *Barassica juncea* [Indian mustard] and *Thlapsi caerulescens* [Alpine pennycress]), they will nonetheless ascribe to other categories on the chemical uptake spectrum—as either accumulators, tolerants, or precipitators. Though they may not figure into site cleanup work, their gene regulation may allow them to sustain themselves despite their relatively high accumulated metal loadings, and even if there should be no contaminant translocation occurring.

Assuming a sufficiency of interested parties invest their energies in this study, information on the toxicological resilience of uncommon test species (but what notably are the common field species) to xenobiotic compounds will be brought forward for the first time. The study may reveal that plants that occur in the field can sustain themselves even where experimentally achieved plant concentrations are higher than would be encountered at a site. The study may drive home the point that what might be construed as compromised biological function (e.g., a reduced photosynthetic rate) frequently goes unnoticed in contaminated environments. To the extent such is true, the information brought forward can focus discussion on discriminating between utilitarian and non-utilitarian criteria for ERA needs.

The study will likely draw in the second-order toxicology concept. To the extent that differences in leaf (chemical) concentrations, and in overall appearance (e.g., height, leaf color, dbh) are observed for the plants of dosed-soils versus controls, ERA practitioners will again be alerted to the essentiality of knowing how much of a difference constitutes a deficit. In an applied context, would it matter if a deciduous shrub grows to an average height of 9 feet in contaminated soils, while the plant in nearby clean soils attains a consistent height of 10–12 feet? That is, would it matter to the minimally stunted plant itself, or to the health and behaviors of animals that might interact with it?

Study #17

Re-Assess *Eisenia Fetida* in Toxicity Testing and Site-Assessment Work

Premise

Earthworm toxicity testing routinely proceeds in support of ERA work for contaminated terrestrial sites. The testing is intended to provide a useful line of evidence for the assessments, either upholding or rejecting a contention that soil contaminants are harming the ecosystem through weakening or killing off worms that are said to form the basis of the food chain. Procedurally, site soils are removed to the laboratory where commercially available earthworms are placed in glass jars of soil. Following test durations of 14 or 28 days, the toxicological endpoints of growth and survival are reviewed for specimens that were placed into either site soils (including dilutions of these) or pristine/reference soils. The earthworm of choice for this work is *Eisenia fetida*, a species with a long list of common names—red worm and manure worm among them—and a species that, too, is potentially quite inappropriate for toxicity testing in support of terrestrial ERA. It is clear that, given their druthers, biologists and ecotoxicologists (and ecological risk assessors as well, it is hoped), would have a worm such as *Lumbricus terrestris*, the common night crawler, be the test species ordinarily used in such ERA work. *L. terrestris* and a good number of related species are anecic, meaning that they construct permanent deep vertical burrows in cool soils, where they feed in decaying organic matter. Thus, *L. terrestris* and its relatives allow for the consideration of contaminants positioned at lower soil horizons, a not uncommon site arrangement. If not for the direct effects that soil contaminants may confer on worms, ecological risk assessors are also wont to examine the role of the earthworm as a contaminant conduit. The interest here is in the worm serving to transfer lower strata site chemicals (that either reside in the worm's ingested gut soil or manifest in living worm tissue) to ground-dwelling receptors (e.g., American woodcock) via the latter's dietary intake. *L. terrestris* is a species that is difficult to raise domestically, and in *E. fetida* having subsequently been taken as a replacement (surrogate?) species, it may well be here that efficacious earthworm toxicity testing (and other well-intentioned earthworm-based study) unravels. *E. fetida*, though easily bred and cultured in the laboratory, is inconveniently an epigeic as opposed an anecic species, living on the soil surface and venturing no lower than 10 inches (25 cm) into the soil column. Its common name, "the manure worm," reflects the species' existence in manure piles. But it is likely that manure piles do not typify the ground surface at contaminated sites. With *Eisenia* confined to the upper 10 inches of soil, the reigning worm test species of choice cannot inform on the chemical transfers to terrestrial vermivores that ecological risk assessors so dwell on. And there are other challenges to *Eisenia* use. While its optimum ambient temperature is in the range of 68° to 77°F (20° to 25°C), the species becomes stressed at 85°F, and can die quickly when temperatures reach above 90°F. Would we, then, expect to find *Eisenia* at contaminated sites where manure piles are not present in the main? What sustainable populations of *Eisenia* could there be at sites where the summer months bring temperatures of 85°F or greater? In an overall sense, what line of evidence can *Eisenia* reasonably provide when it is so lacking in the area of site earthworm surrogacy?

This study is directed at ascertaining the efficacy of standard contaminated site soil toxicity testing with *Eisenia fetida*. Specific study elements include exploring the potential for *Eisenia fetida* presence at contaminated sites and documenting actual sightings of this species at these locations; conducting comparative *Lumbricus* spp. and *Eisenia fetida* toxicity testing; reviewing earthworm population data for contaminated sites and particularly where desktop assessments indicate their seeming inability to be supported; and the development of improved earthworm toxicity assessment schemes.

Study Guidelines

1. For terrestrial sites that had *E. fetida*-based toxicity testing run as part of the ERA work that was conducted, an effort should be made to document if/that manure piles existed where soil samples were collected for any aspect of ERA review. Reconstructing site conditions from decades back need not be as difficult or laborious as one might think, for published RI-level documentation would necessarily have reported such site features if they were prominent. Conventionally, RIs provide rather thorough "Site Description" and "Ecological Setting" narratives; while a singular manure pile measuring just a few feet by a few feet might not be

described, where manure piles were prominent, we could imagine text on the order of: "The site is marked by some two dozen manure piles that constitute some 40–45% of its surface." As for sites still undergoing ERA work as part of some regulatory environmental restoration (or equivalent) program, the opportunity exists to directly observe the site layout, and thereby document the presence of manure piles constituting a significant portion of a contaminated area.

2. For ongoing ERA investigations at terrestrial sites and for historically reviewed terrestrial sites, this study encourages gathering earthworm species lists. It is understood that, for all the efforts expended in the area of species identification at contaminated sites, typically it is only for birds, mammals, reptiles, and amphibians that species presence information is reported. Even then, the information provided is conventionally of two types—"species known to be present (onsite or in the vicinity)" and "species thought to be present." Researchers should be prepared to find that, for many sites, earthworm species identification information and earthworm censusing data is fully lacking. (The researcher should appreciate the complication raised by absent earthworm community descriptions. Putting aside the veritable non-contestable claim that *E. fetida* is a non-adaptive and inapplicable test species, the situation begs certain questions: Why did *E. fetida* toxicity testing proceed when no one bothered to establish if a given site was lacking earthworms? Why was earthworm toxicity testing of any order conducted? Assuming earthworms *were* observed while sampling soil in support of site ERA work, why weren't these identified to species?)

3. For sites with ongoing ERA investigations, an effort should be made to apply commercially available *E. fetida* to the ground surface at well-marked areas in contaminated zones. After some weeks, a subsequent effort would involve trying to recapture these worms and/or to document the extent to which the added worms had either acclimated or had not been able to survive (in true site soil, which we have seen departs radically from manure).

4. Securing a fuller understanding of the topical area, enthused researchers should set about to devise an earthworm toxicity testing scheme with one or more anecic species. Understanding that earthworm toxicity testing is only relevant where it has first been established that a contaminated site limits or entirely precludes earthworms, the long-range goal would be to conduct testing with species that are otherwise associated with a site. A necessary component of the developed testing would be directed at identifying species sensitivity differences of *E. fetida* and the anecic worm species of the to-be-developed testing scheme(s).

Study Outcomes and Applications Thereof

The information to be gleaned from the suggested work bears the potential to spark a re-thinking of the role of earthworm testing in support of ERA in a holistic way. Potentially we stand to definitively learn that *Eisenia fetida* is an inappropriate test species. Its worthiness will be challenged (a) should a survey reveal that manure piles are not a predominant landscape feature of sites, (b) should *Eisenia* applied in the outdoors (even at pristine locations) fail to survive, or (c) *Eisenia* not be found to naturally occur in manure-free contaminated locations. The gathered data will likely lead to certain ERA reform considerations: why does earthworm toxicity testing so often proceed when a site's earthworm community has not (first) been characterized? How can a toxicity test species of choice attain such a status if, when placing the worm at a contaminated site, it is forced to live (in the ambient condition) outside of its thermal tolerance limits? What was the toxicity test approval-granting body thinking when it moved for indoctrinating *E. fetida* as an ERA tool that could supply helpful ERA line of evidence information? Where Eisenia is shown to be an inappropriate test species, standard ERA practice for terrestrial sites would need to adopt a stronger justification for proceeding with the conducting of earthworm toxicity testing. Earthworm censusing that could establish that populations are depauperate or altogether absent would become a prerequisite for conducting earthworm testing of any sort. Further, where the failings of *E. fetida* testing would be exposed, only developed anecic worm toxicity testing (ideally with a species that actually occurs at a site if interest) would be sanctioned from this point forward.

Study #18

Monitor Electrocardiograms (EKGs) of Avians (and Other Species) at Contaminated Sites

Premise

The reality is that sick-looking birds or mammals at contaminated Superfund-type sites are not described in ERAs. There are several different ways to account for the reality. First, it could well be that as variable as sites are with regard to the contaminant suites they bear, none of these present health challenges to these two animal groups. (Without a doubt, this notion runs contrary to the way some ERA practitioners, regulators primarily among them, think.) Second, it may be that ecological risk assessors, and the wildlife biologists they rely upon, spend insufficient time in the field to allow for the observation of direct health impacts that might be present. Third, it could be that contaminant-posed animal sickness that leads to death surely occurs, but not to a degree that is noticeable. Aside from not upsetting the local ecology, an occasional animal that succumbs won't be evident to anyone, and the reality too is that carcasses disappear from the scene rapidly even when exposure to improperly placed chemicals is not the cause of an animal's death. A fourth reason: vast numbers of sites are frightfully small, and to the point that there are no sizeable populations of the species that truly concern us regularly occupying them. Given the foregoing, the point to secure is that if we cannot document overt signs of compromised health, let alone happen by a sufficiency of animals to make ecological health investigation worthwhile altogether, a shift in ecological site review is in order. Astute ERA practitioners will recognize that the shift cannot involve toxicity testing, for investigations of that type are simply too divorced from a site's ecological doings to yield meaningful information; we necessarily have the need to know about the health of the animals that live at contaminated sites, and so we must draw closer to them. While they may not be dying or moribund, they might be unhealthy in other and more subtle ways. We should, then, seek to know if site animals are physiologically functioning as they should or if, due to their contaminant exposures (as opposed to the effects of aging), they function at the subpar. Thus, the interest here is not with chemical body burdens, but rather on how vital organ systems run. As an example, the pulmonary system could be investigated, where the researcher would be looking to assign a fitness ranking based perhaps on (age-adjusted) lung capacity and/or lung inspiration volume. Importantly, all-critical fitness rankings to be assigned to site specimens when using this approach will necessitate a knowledge of the evaluated physiological function norm; simple comparative assessments of site- and reference-location animals (where, realistically, the former present with lesser performance figures) will not be serviceable. This writing recalls Study #15, which anticipates and welcomes comparative assessments for contaminated and non-contaminated areas. The information to be gleaned from those studies certainly has its worth; it can most definitely add to the body of knowledge on field animal toxicological response, and, more fundamentally, reveal how plastic a given physiological parameter may be.

Due to the intricacies involved, this study goes beyond the comparative assessments that seek out somatic differences between exposed and non-exposed animals. It specifically targets the cardiovascular system, recognizing its centrality in overall health. The electrocardiogram (EKG) is defined as a graphic record of sequential, electrical depolarization-repolarization patterns of the heart. Enthused researchers must be in touch with the select purpose behind the undertaking, namely that of trying to understand why (bird and mammal) ecological receptors can continue to populate contaminated sites despite assessments that predict the very worst for them. Researchers must be comfortable with breaking off from the conventional ERA process (that, pursuant to HQ assessments, might/would invest energies to determine if certain predicted effects actually appear in field animals) to focus on cardiovascular health alone, i.e., an endpoint that will likely bear no relation whatsoever to the modes of action of a given site's contaminant suite. The approach to study thinking should be as follows: "With the contamination present at the site, organism health stands to be compromised. Based on documented animal presence at the site however, the anticipated health compromise appears to be lacking. Let us, then, look a little closer, beginning with a review of a physiological function that can ill afford to be unnaturally offset, namely cardiovascular health." Importantly with regard to birds, the skids are greased for study; more and more veterinarians in recent year have incorporated into their practices a unique brand of specialized testing and diagnosis, the recording and interpretation of the EKGs of pet birds. It is critical to note that, despite its clinical applicability, EKGs have received relatively little attention from those working with wild birds. Perchance this is due to the scarcity of electrocardiographic reference values in birds. In the face of such data gaps, the open literature (some of it dating back more than five decades) nevertheless reports stressor-cardiac effect associations. These include thiamine deficiency-induced sinus arrhytmias and ventricular premature contractions, ST segment decreases in chickens, and bradycardia and tall T waves in ducks with hyperkalemia. While the matter of lacking vetted EKG reference values for wild birds is real, the present study can be launched nevertheless. Bird EKGs have the capability to detect serious conditions in wild birds as they already reliably do for pet birds.

Study Guidelines

1. Compile a list of contaminated terrestrial sites, as yet not remediated, for which unacceptable avian HQs have been computed. Review bird species lists for the sites, identifying one or more species that, based on documented behaviors and spatial dynamics (feeding and home ranges relative to the size of the contaminated parcel), display a high degree of site fidelity (which researchers should quantify and keep a record of). For each qualifying contaminated site, identify one or more suitable reference locations. These are defined as those that satisfy two essential criteria. They must first support a bird species to be investigated at its paired contaminated site. Second, they must be located as near to study sites as possible, but sufficiently far enough away from them such that birds who predominantly occupy the reference location(s) would only on highly infrequent occasions, if at all, appear at the linked contaminated site(s).

2. Capture birds through any means (e.g., mist netting), endeavoring wherever possible to minimize stressing them. Document the latitude/longitude location of each bird's capture (see guideline 6). A minimum of five species representatives from each of a given contaminated site and its paired reference location will be needed to constitute a valid contributing element for the study. Birds of a given species need not be captured on the same day. It follows, then, that birds should not be housed to await being heart-monitored, only after the requisite number of species representatives has been assembled. Study enthusiasts will understand that, although some birds may have had their EKGs taken, they may not end up being included in the study should the required number of conspecifics not be assembled.

3. Convey birds to the laboratory, and appropriately house them. (Socially house birds, if at all possible, to abide by the recommendations and/or

requirements of the most recent laboratory animal care and use manuals). Research and provide a suitable (species-specific) acclimation period before recording any vital information that might bear on a bird's heart function.

4. Gently restrain birds, placing them into appropriate harnesses or equivalents, and affix electrodes as necessary. Record the EKG; collect traces until there is at least one of sufficient duration that reflects the animal at its resting state (i.e., free of stress), to be reliably reviewed by a specialist. Recognize that certain species may not prove to be suitable for study owing to behaviors whereby the animal cannot be freed of agitation when restrained, to allow for the recording of usable EKG traces. Under no circumstances are birds to be administered sedatives or similarly acting drugs in order to induce a calmative state.

5. Document all instances of erratic or irregular EKGs (e.g., those showing ventricular arrhythmias, bradycardia, etc.) and evidence that an animal might have experienced a past cardiac event.

6. After EKGs are recorded, release birds at their point of capture.

7. Endeavor to establish linkages (correlations) of site chemicals and observed heart effects. Endeavor to collect chemical concentrations in soil, diet items, (bird) blood, and other available tissues (e.g., feathers, hair).

8. Endeavor to record the EKGs of mammals that appear to be amenable to study. Apply guidelines 3 through 6 in that regard. As with the bird work, the study component demanding the greatest attention and care will seemingly be that of restraining test subjects with an appropriate harnessing apparatus.

Study Outcomes and Applications Thereof

The primary merit of engaging with the subject study is the potential for discovering an alternative and valuable means of assessing the health of contaminated site receptors. The premise began with a brief review of four possible ways to explain the absence of documented debilitating, if not lethal effects in birds and mammals, the only two phylogenetic groups assessed in terrestrial settings. In a large way, study findings should nurture the discovery of the one or perhaps two best resolutions to the matter. To the extent that contaminated sites simply do not house enough animals—because sites are so relatively small—such that biologically and ecologically debilitating effects can reign, the study bears the capacity to put enthused researchers in touch with this phenomenon. They will have no choice but to come to grips with the key matter of ERA purpose, being moved to ask for the common case, if a true need exists for the ERA field altogether. Realistically, researchers might not be thinking that capturing just five animals of one type (e.g., grouse, vireo, cottontail, raccoon) will present its difficulties and even at non-contaminated reference locations. Such oversight can stem from a non-familiarity with spatial dynamics, or from being distracted by the novel/atypical study design elements here, such as fashioning one-of-a-kind appropriate harnesses and electrodes. To the extent that the study reveals that birds and mammals have heart damage, follow-on questions are of only two kinds: What site attribute(s) specifically caused the damage? and Does the observed heart damage effectively matter? These questions will automatically fall away should EKGs in reference-location animals reveal cardiac conditions or evidence of past significant events in roughly the same frequencies as occur in the site animals. Perhaps we stand to learn that, as with aging humans (and despite their maintenance of a proper body weight, diet, and exercise regimen), heart irregularities and other manifestations of heart disease (e.g., high blood pressure) arise as a matter of course. Where cardiac effects are present in site animals to a notably greater extent than in reference location animals, the meaningfulness of this will have to be decided. This could commence with a consideration of wild species longevity. The older that animals with heart effects are, the more of a demonstration there will be that observed effects are inconsequential. (This recalls our understanding of cancer vis-à-vis ecological receptors, and specifically why cancer is not held to be a toxicologically relevant endpoint for them. We know that tumors only arise in non-human species at senescence after any and all reproductive contributions have been made.) To the extent that cardiac effects are found to correlate with chemical suites, the study will have given rise to a potentially lucrative (in a risk assessment context) evaluative scheme. In that case, the development of a library of species-specific EKGs, both in contaminated and non-contaminated settings, could prove invaluable for future ERA work. Essentially, we would develop knowledge of the kinds of cardiac system effects (i.e., aberrations) one could expect to see in animals that frequent sites with certain contaminants, and a sense of those aberrations that are/are not indicative of a problem.

Study #19

Dose Test Animals with Aged and Environmentally Weathered Chemicals

Premise

Irrefutably, the chemicals administered to test animals in the laboratory for the purpose of deriving TRVs in support of HQ-based assessments are brand new. This is most understandable, and, in one regard at least, is probably a responsible practice. With very minor exception, test chemical costs are minute, paling in comparison to the costs of purchasing, housing, and ensuring the veterinary care of test animals (e.g., rats, quail). Test chemical costs also pale in comparison to research staff salaries and critically needed specialized equipment. Chances are, then, that even if researchers should have quantities of a test chemical on hand from prior use, they will "start off clean," opting to order a new lot of a test compound from a chemical supplier. It is here, though, that a formidable testing complication arises. For routine dosing studies, some might think it admirable that brand new batches of test chemicals with near absolute purities, and that have likely been tightly stoppered and light-shielded from the elements, are the ones employed. The reality, though, is that the chemicals to which ecological receptors are exposed at contaminated sites have had unlimited exposure to the elements and are anything but pure. (Interested parties should recall that sites that submit to ERAs are always three or more decades old.) Reasonably, the unadulterated and unreacted chemicals used in TRV development bear a maximum toxicological potency, and, reasonably too, the environmental weathering of chemicals bound up in soil has acted to significantly attenuate the latter's potency. The seeming manifestation of this arrangement, something that uncertainty sections of ERAs continually fail to discuss (probably due to oversight), is that the TRVs to emerge from laboratory testing are probably excessively stringent. The excessive stringency has its consequences; many or perhaps most failing HQs are likely not so, and, in turn, recommendations to conduct further study or to have sites advance to the cleanup stage are, respectively, likely unnecessary and unfounded. This study is not simply recommending that ERAs make it their business to tighten up their uncertainty sections, highlighting the great potential for exaggerated HQs to be computed. This study is primarily recommending that inroads be made to stem the tide of generating NOAELS, LOAELs, and TRVs that are not only undoubtedly overly stringent, but that, in principle, do not align with the field condition being assessed.

This study recalls that TRV usage is only applicable to terrestrial systems, and specifically for bird and mammal evaluations where the singular chemical uptake pathway assessed is ingestion. In brief, chemicals to be administered to test animals will first need to be aged and weathered. Following this, creative means will need to be devised for amending animal diets with the aged and weathered test chemicals. To the extent that soil is to be fed to test animals, interested parties should note that incidental soil ingestion accounts for some 2% of a mouse's diet and some 10% of a bird's diet. Interested parties are reminded as well that chemical forms and states (e.g., valence) change while weathering, often substantially. By way of example, a specific metal salt amended to soil that is subsequently wetted (whether under experimental conditions in the laboratory, or as occurs naturally with incident rain), will, perhaps within just minutes, cease to exist as that very salt (see guideline 4).

Study Guidelines

1. This study is again one that creates vast participation and collaboration opportunities for committed researchers. A consolidation of the findings of researchers working independently can make for powerful contributions to ERA science. The scope of the study is broad, extending to all chemicals for which one or more dietary-based TRVs exist, for birds, mammals, or both groups.

2. The study's primary objective is to recreate the dosing arrangement of the studies that have given rise to vetted TRVs presently in use. In dosing test animals with "now-weathered" chemicals, ideally the same test species and strain should be used, and the mode of administration, exposure duration, and measures of effect should all match those of the studies being duplicated. An effort should be made to select studies that yielded a NOAEL and/or LOAEL for a reproductive endpoint (e.g., litter size, pup survival) as either a stand-alone effect or as one of several effects monitored.

3. In every case, the chemical form purchased should be the one that was used in the toxicity testing that gave rise to the vetted TRV. See next guideline.

4. For each chemical, determine the concentration in soil (in mg/kg) that coincides with the (daily) dose administered in the adjusted diet of the vetted study to be duplicated. If (encapsulated) soil is not to be administered, devise a method of administering aged/weathered soil through the diet, where the test animal will receive the same mg equivalents as were administered in the vetted study. Recognize that the chemical-in-soil concentrations sought may need to be (considerably) higher than those to ultimately be arranged in the laboratory diet. (Reclaiming the post-weathered chemical from the soil matrix, reconstituting it, incorporating it into a slurry if necessary, and other procedural measures may contribute to compound losses. Additionally, en route to amending animal food such that it bears chemical concentrations akin to those of vetted studies, initial soil concentrations may need to be notably higher.) Vetted studies will need to be thoroughly reviewed to understand with certainty what "oral in diet," as the exposure route, specifically involved. Because adjustments of soil concentrations to "oral in water" and "oral gavage in water" are anticipated to be complex and unreliable, restrict study efforts to only "oral in diet" studies for birds and mammals.

5. Endeavor to use at least two soils that notably differ in any or all of the following characteristics: particle size distribution, pH, water holding capacity, cation exchange capacity, organic matter content. Establish the amended soils in very large pots that are to be sequestered in the out-of-doors with no unnatural shielding (e.g., roofing; a layer of plastic mesh, though, can be placed atop the pots to minimize debris falling in, and to remove animal interferences). Allow for drainage in the pots, understanding that some compound loss via percolation through the soil column will be unavoidable. Arrive at desired plant pot concentrations to be aged/weathered through repeated adjustments involving laboratory analysis of a representative soil aliquot, subsequent addition of soil or chemical, thorough mixing, and aliquot removal for laboratory analysis once again. Allow a variance of ±10% for the intended concentrations to be achieved, recognizing that, even with excessive mixing, arriving at perfect homogeneity is unlikely. Ensure that soils to be amended are contaminant-free. With regard to metals to be tested, analyze soils at the outset to know the naturally occurring concentration. Such will, in turn, provide indication of how much additional metal will need to be spiked.

6. A second chemical aging/weathering procedure is suggested, as follows. Place a known/measured quantity of a chemical, which will initially often be in powdered form, into a clean all-glass (i.e., inert-surface) container with high walls (e.g., a 30-gallon aquarium). Secure the container outdoors (perhaps on a building roof), where there will be no shielding from above. As with the pots containing amended soils

(guideline 5), a plastic mesh may be used to restrict debris from falling into the containers. Allow amended-soil pots and glass containers to weather outdoors for a full calendar year. (See guideline 12.) Reclaim the pots and containers and allow soils and container contents to completely dry. Fans may be used to hasten the drying where it is clear that such will not contribute to chemical loss.

7. Analyze one or more representative soil aliquots for the post-weathering year concentration. Return to guideline 4 and its imperative to administer weathered chemical to test animals.

8. Fully contents-dried glass containers will need to have their chemical holdings, that may be invisible, collected (via some reconstitution procedure) in a manner that does not degrade compound toxicity.

9. Conduct the necessary animal dosing studies, adhering as closely as possible to the tenets of the studies supporting the vetted TRVs (with, ideally, the only difference being the use of now-weathered chemical).

10. For all endpoints, and with a particular focus on reproductive ones, compare NOAELS and LOAELs for the vetted studies and the new testing. Ensure that the computations of the vetted studies are those used in establishing the new NOAELs and LOAELs. Identify all instances where a weathered chemical is observed to have a higher threshold for toxicity, and tabularize the degree to which vetted NOAELs and LOAELs are unnecessarily stringent (i.e., biased low). Revisit completed ERAs, i.e., those that employed the vetted TRVs we speak of. Identify all instances where ecological-based cleanups were recommended or actualized, but where HQs would be acceptable if newly crafted, weathered chemical-based TRVs had been used in the assessment scheme.

11. Particularly where vetted TRVs are found often enough to be overly stringent, interested parties are encouraged to develop NOAELs or LOAELs where the experimental weathering period extends to two years or beyond.

Study Outcomes and Applications Thereof

If nothing else, for a great many chemicals sequestered in soil for 30 or more years, bioavailability is lessened. Incidentally ingesting aged and weathered contaminated soil, which for some birds and mammals is the only means through which chemical exposure occurs (as through earthworm consumption), stands to pose a significantly lesser toxic challenge than most realize. Either because plants are often naturally poor bioaccumulators (i.e., root uptake and chemical translocation from the roots to above-ground plant parts is not a dominant phenomenon) or because chemicals over a period of decades become tightly bound to soil particles to a point that plants do not incorporate them, plant consumption does not stand as a chemical threat to herbivorous species. Offering weathered chemicals to test animals, particularly when the chemicals occur in the very matrix in which they weathered, much more closely simulates the chemical exposures of ecological receptors than do the dosing regimens of vetted TRV-generating studies. An anticipated finding is that newly crafted NOAELs and LOAELs for an array of chemicals will be discovered to be significantly greater than currently available ones. Early indications that vetted TRVs are unduly stringent should trigger an exhaustive chemical toxicity assessment review in the form of a mandatory retesting program, whereby common test species are orally dosed with weathered chemicals. To the extent that weathered-chemical NOAELs and LOAELs are shown to be higher than vetted ones, and (consequently) computed weathered-chemical HQs to be of lower magnitude than those of completed assessments, the potential will be born for weathered-chemical animal dosing to explain how toxicologically impossible HQs come about.

Study #20

Assess Resistance to Toxic Effect Expression through Dosing Chemically Pre-Conditioned Animals

Premise

One of the weakest features of the toxicological testing that supports the TRVs used in terrestrial ERA is that test subjects pressed into action have had no prior exposure to test chemicals. Unfortunately, that arrangement is anything but the case for the receptors in the wild that concern us. Site receptors live in a contaminated environment and consume contaminants daily. For some, the only exceptions to their ongoing daily exposures are "time off" for migration to warmer climes in the winter, or for living in a torpid (i.e., non-feeding) state for some weeks or months at a time, while still residing within a site's boundaries. Site receptors, then, are quite familiar with contamination, and contaminants can often be detected in their tissues (e.g., fur, bone, blood). Further, detections of site contamination in animal tissue can supply any corroboration needed that certain specimens culled from the field are site representative. Importantly, too, there is every reason to suspect that an animal's somatic chemical stores reflect not only what a given field specimen has consumed or contacted, but what has been maternally transferred to it. As has often been noted in this compendium, test animals that hail from animal breeders share no part of a chemical-exposure history we so wish they had. More to the point, while test animals without a contaminant exposure history might work perfectly well in support of HHRA, ERA has no interest whatsoever in the toxicological responses of animals of this kind. Consider that it hardly happens that chemically free animals find themselves suddenly, for the first time, dealing with contaminated media. Reasonably, toxicological effects are more easily elicited in laboratory-bred and reared animals than in animals that have experienced prior chemical exposures. For the ramifications it might hold for TRV development and application, it would be prudent to try and illustrate this.

The subject study encourages enthusiastic individuals, preferably those with a background in traditional ecotoxicology (as in hands-on animal dosing), to apply an experimental protocol such as the one suggested here. The goal is to engineer test animals (mammals and birds) that mimic, in a most critical area (chemical sensitization), the ecological receptors to be assessed (albeit weakly, following HQ methodology) at contaminated terrestrial sites. Ostensibly, these engineered animals should be more appropriate for use in TRV development than those presently used. Realistically, this is the most simplistic of the compendium's laboratory studies. That said, it is only fair to acknowledge that the study has its share of tedium born of a repetitive-task nature. This however, should be a small price to pay for the opportunity to better understand chemical sensitivities and animal resilience (resistance to toxic effect development) for ecological receptors living in contaminated environs.

Study Guidelines

1. Cost considerations should be the only obstacle to overcome en route to having this study materialize. The high costs are driven by the animal care being rendered, specifically the expenditures for prolonged animal housing and the necessary veterinary oversight that the project entails. In the best of all possible worlds, the EPA, as the essential regulatory body vis-à-vis implementation of TRV-based ecological assessment, should champion the initiative that stands to reveal whether or not TRVs in use are accurate or, unfairly, over-conservative.

2. As with a number of other studies of the compendium, there are truly limitless opportunities for enthusiasts to involve themselves with the work. Any chemical with a TRV should be tested, and multiple common test species can be appropriate for use. (Note: the study guidelines pertain to a conventional rodent testing sequence. Adaptation of the study design to facilitate the study of birds [e.g., pigeon, quail] should be straightforward. The additional step-wise work in the case of birds [e.g., candling, incubating, and hatching eggs, etc.] is acknowledged. See next guideline.)

3. Although not absolutely required, it is suggested that study participants endeavor to replicate TRV supporting studies in terms of test species, chemical form used, and toxicological endpoint to be assessed. At the same time, with the interest in bringing the best information forward,

participants should try to conduct their work on two different species (e.g., a house mouse and a white rat) where a more truthful mammalian TRV is being sought). Study durations would ideally match those of present TRV-supporting studies. It is recognized that two-year study costs would be excessive, and might not allow for the necessary successful matings of dosed animals. Since 90-day exposure studies are quite the norm, and fully defensible as chronic in design, consider running these, and certainly where 90 days was the exposure duration of the TRV-supporting study being replicated.

4. Each study should begin with three animals groups, as two to be identically dosed with the test chemical (for 90 days), and the third group, serving as controls, to be dosed with the test chemical vehicle only. After the dosing phase, euthanize the controls and one of the dosed animal groups. Proceed to assess the sublethal endpoint(s) of interest (e.g., reproduction, perhaps with litter size the specific focus). Note: it is likely that the study to be replicated only tested with one sex. Where young production (as successful matings/number of litters, litter size) is to be monitored, ensure that dosed animals are mated to chemically free animals (e.g., proven breeder males). Depending on circumstance, derive, at a minimum, a chronic NOAEL or a chronic LOAEL (and perhaps both of these). Retain the derived toxicity values (see guideline 6).

5. Breed the second/remaining dosed animal group to normal mates. (Ideally both males and females were dosed.) Monitor pregnancies to parturition, count and weigh pups at birth, and house pups with the dam until weaning. Continue to dose pups, group-housed by sex, until 90 days old. Breed this cohort to normal (i.e., non-exposed) mates as at the start of this guideline, and follow through with weaning and dosing as described above. Endeavor to repeat the course of dosing, mating, and rearing offspring through a minimum of five and ideally ten generations. (Breeding 90-day dosed animals for more than ten successive generations would certainly not be discouraged.)

6. Proceed to derive a chronic NOAEL, chronic LOAEL, or both, by dosing sexually mature control animals and the reared and sexually mature offspring of the last mating of dosed animals. (Alternatively, the toxicity data of the control animals of guideline 4 can be used, assuming, of course, that the mating protocol implemented for the "successive generations-dosing and breeding" animals is identical to that used earlier.) Compare the toxicity values with those derived after a sole 90-day exposure (as described at the end of guideline 4).

7. Tabularize all cases of chemically preconditioned animals having greater thresholds-for-effect than conventionally tested animals, presenting for each the degree to which traditional TRVs exaggerate toxicity.

Study Outcomes and Applications Thereof

Many features of conventional laboratory animal toxicity testing (e.g., a fixed light-dark cycle, constant ambient temperature throughout the dosing phase) are unavoidable. It is unfortunate that, for virtually every setting (and not just in support ERA through TRV derivation), imposed environmental constraints are never acknowledged or discussed. How helpful it would be if ERA uncertainty sections owned up to the task. Acknowledging that a LOAEL used in an ERA derived from a study where room temperature never deviated by more than one degree over weeks of testing, and that the evaluated receptor's natural environment varies by 30 or more degrees daily, would paint TRVs in a new light. If uncertainty sections noted that TRVs derived from (mere) cohort studies, and that wild receptors are exposed for tens of generations and decades of time, movers and shakers would seize on investigating this difference. For now, it doesn't appear that anyone, whether harboring a purely academic interest, or an ERA-focused one, has decried the complication. Another view asks how it is thinkable that scientists have managed to look past the disparity. In any event, free for the taking is the chance to discover if generations of chemical exposure confer a resilience to toxicity endpoint elicitation. Should this be the case for even a single chemical, one would imagine that other interested parties would want to follow suit to flush out the toxicity question. Here it is worthwhile noting that the concept of continuous breeding is not a new one; in support of tracking impairments to rodent reproductive via chemical exposures, a hefty continuous breeding database exists.

Should NOAELs and LOAELs be discovered to be higher set than what present toxicity databases purport (i.e., chemically preconditioned animals display greater thresholds-for-effect than do animals of cohort studies), there could be great ERA ramifications. (Finding that chemically preconditioned animals are less resistant to chemicals effects than are cohort study animals is not an expectation.) At a minimum, uncertainty sections would come to routinely acknowledge the phenomenon. One or more findings of enhanced chemical resiliency would be expected to translate into weaker cases for cleanups being put forth, and for a strong push to replace all NOAELs and LOAELs with those based on testing with chemically preconditioned animals.

Study #21

Develop a Valid and Reliable Environmental Residue Effects Database for Freshwater Fish

Premise

To a greater or lesser extent, all ecological receptors contacting contaminated environmental media develop a chemical body burden. Fish undoubtedly bear the greatest proclivity for doing so and for the most obvious of reasons, namely that the organism, in its natural state, exists fully bathed in its surrounding medium, facilitating chemical uptake that (in addition to dietary inputs) can and does occur passively. Regardless of the body compartment(s) amassing the chemical residues, there are two distinct concerns articulated in ERAs and ERA guidance. There is the concern that the assimilated toxic chemicals will either impact normal biological function or cause premature death of the organism bearing the toxins. The second concern is over residue transfers occurring through the aquatic food chain, and the health effects that these stand to potentially trigger in piscivores. It is unfortunate that ERA science has not seriously investigated the transfer of chemical residues of smaller fish to larger fish, although designing studies to understand this phenomenon are seemingly easy enough to assemble. Those defending the absent information can reasonably only put forth one argument. ERA for fish does not involve a food ingestion model that (its weaknesses aside) would parallel what is ordinarily applied in terrestrial work. Importantly, where the health of birds and mammals is a concern because these animals have diets that include fish, there is no barrier to knowing of the true (as opposed to modeled) body burdens of fish that would serve as prey. It is easy enough to catch some site fish to fittingly enable the standard intake equation for bird and mammal receptors-of-concern.

This study is directed at the first and more elemental phenomenon. Since ERA claims to be concerned with the welfare of aquatic ecological resources, and fish indisputably assimilate chemical residues, there is a distinct need to know if residues are harmful to the fish that bear them.

This matter needs to be assessed without any rumination about carry-over effects, i.e., the potential for species higher up in the food chain to be chemically overcome. Importantly, to summarily conclude that fish are unhealthy because they bear chemical residues of any sort is to side-step ERA's mission. Assuming that a fish or any other ecological receptor is health-challenged simply because it lives with, or its body stores, contamination amounts to no more than conjecture, and the practice reflects poorly on ERA practitioners; why aren't they motivated enough to investigate just what body residues are harmful? "Guilt-by-association" (Tannenbaum, 2013) is not an acceptable way when the investigative means exist to definitively know what chemical residues portend. At this juncture it is worth noting a common collapsed argument of ecological risk assessors. If fish are being undone—poisoned to the point that they can only meagerly reproduce, if they can survive at all—there is no room left to trump the playing card of ecological havoc occurring through the linked "contaminant-transfer-through-the-food chain" supposition. ERA, then, can't have it both ways. Potentially, chemical residues in fish can reduce fecundity, render fish less hardy, produce stunting, or shorten the lifespan. These and other toxicological endpoints need to be investigated, and importantly without relying on modeling or the responses of common aquatic test species approaching only two inches in length (e.g., mosquitofish, fathead minnow), and exposed in limiting artificial laboratory settings. The subject study invites multiple party interests to investigate a range of fish species and a range of chemicals.

Study Guidelines—Laboratory

1. Arrange to have testing conducted using large indoor (or outdoor) pools, and with fish species that are minimally 6–8 inches in length (e.g., black crappie; *Pomoxis nigromaculatus*). Pools for both fish to be chemically exposed and to serve as controls should be able to support 2–3 dozen fish. Intend to test at least two fish species. Fish may be segregated in pools by species or mixed within pools. See guideline 2 regarding the feeding of the fish.

2. Provide a species-appropriate sediment layer 4 or more inches thick, preferably by transporting sediment from a non-contaminated waterbody (for, in this way, it will contain the other life forms that normally populate a natural aquatic system [e.g., a lake]). Anticipate the dietary needs of the test fish species. These may range from plankton and minuscule crustaceans to small fish (e.g., shad, minnows). Where feeding complications might arise with two fish species in a pool, test only one species at a time.

For indoor pools, supply a variable light/dark cycle that approximates day-length changes occurring over a goodly portion of the calendar.

3. Test one chemical at a time. For each (organic and inorganic) chemical to be tested, determine the highest sediment concentration recorded at a Superfund (or equivalent) site. Over a series of pools, apply a chemical (e.g., cadmium) to the water in a form in which it has been documented as having been released at one or more contaminated sites, with the goal of achieving the researched highest sediment concentration. Allow several weeks for the contaminant to partition to the sediment and for system equilibrium (between water column and sediment) to be established. (Under no circumstances are chemical residues in fish to be achieved through artificial means, such as by deliberate injection to the animal. Under no circumstances should efforts be made to achieve sediment chemical concentrations that exceed the highest of those documented for hazardous waste sites.)

4. Review the open literature, to include Superfund and other contaminated site histories and ERAs, to identify highest chemical-specific fish tissue concentrations.

5. Add fish to the pools. Observe fish health (growth, survival, opercular rate, presence of fin rot or external lesions) and behavior (swimming, activity level, schooling, feeding) for several months. For each test chemical, secure from animal physiologists an estimate of the time needed for a maximum body residue to amass. After that estimated time, remove some fish and have whole bodies and livers analyzed for chemical residue levels. Render as invalid pool treatments that gave rise to chemical-specific body residues that exceed those noted in guideline 4.

6. Where desired residues have been achieved, endeavor, to the extent practicable, to mate fish and to track reproductive capability/success.

7. Endeavor to develop neurological function testing capability for fish, perhaps by adapting established testing protocols for earthworms. (Briefly, pursuant to having contaminated jarred-soil exposures, earthworms are placed onto a circuit board, with the ventral side straddling one or more pairs of thin, uniformly and closely gapped [perhaps one millimeter separation] parallel wires, where the earthworm body completes an electrical circuit sensitive to changes in electrical resistance. Subsequent to the application of gentle tactile stimuli to the worm's body wall, electrical responses [such as delays in nerve wave response or decreases in nerve fire amplitude] are recorded.) Quantifiable neurological function changes are worthy of monitoring because they signify animal impairment.

8. To the extent that health-screening measures indicate compromised biological or physiological function in chemically exposed fish, tabularize the following: species, age and size; chemical concentration in whole body and liver; shifted biological function(s); and relative degree of

(assumed) impairment from the norm. Barring exceptions, ensure that the reporting acknowledges the non-existence of thresholds-for-effect for any observed shifted physiological measures.

Study Guidelines—Field and Laboratory

1. Collect larger fish (of 6–8 inches in length) from contaminated and nearby matched reference waterbodies. Verify that fish from the former bear chemical residues and record the whole-body and liver concentrations.
2. Comparatively evaluate fish health (size/length, opercular rate, presence of fin rot or external lesions) and behavior (swimming ability, activity level, feeding). Tabularize the information as in study guideline (lab) 8 above.

Study Outcomes and Applications Thereof

Without clear evidence to support their contention, ERA practitioners often assert that animals bearing chemical body burdens are health-challenged. The contention is in line with a much broader feeling among this professional group, namely that wherever environmental media bear contamination, all those ecological receptors that contact these are operating at a deficit. Although experimental design allowances are unavoidable vis-à-vis the subject study, it is nevertheless arranged to arrive at a truer understanding of the biological implications of fish with unnatural chemical tissue residues. Potentially, the findings will show that the body residues are inconsequential, and that fish with and without residues are indistinguishable from one another. Such information will reinforce the point that just because contamination is encountered, a biological or ecological difficulty need not exist. In that case where health effects are not observed in contaminated fish—either those made to develop a body residue, or those culled from contaminated waters—study participants and those reading about such findings need to be primed for follow-on reprisal. Those who raise the specter of the potential for human health effects to nevertheless develop from the consumption of chemically tainted fish will have blown past the study's purpose, that of looking for *ecological* impacts only. In that case where contaminated fish appear and swim accordingly, and fare as well as do controls, the study will have indirectly and through *de facto* means, championed a related and vital point; we can know of the wellbeing or sickness of an organism from its exterior. Further, identifying biochemical differences between chemically exposed and chemically free organisms, where thresholds-for-effect do not exist for these, has no relevance to health assessment. The study design deliberately avoids the inclusion of enzymatic analysis and related testing because second-order toxicology for endpoints of this nature do not exist (e.g., Does it matter to contaminated fish that their Cytochrome P450 levels are notably different from those of paired non-contaminated fish?). A final intent of the study to potentially be borne out is that there are no ERA gains to be had where the science takes the "easy way out." We are not securing a proper and defensible read on the potential for fish to incur chemical effects if we simply place into aquaria for four weeks, miniature fish that never experienced water-column or sediment contamination before, and that were selected for use because they are easy to rear and maintain.

Reference

Tannenbaum, L.V. 2013. *Alternative Ecological Risk Assessment: An Innovative Approach to Understanding Ecological Assessments for Contaminated Sites.* Sussex, UK: Wiley Blackwell.

Study #22

Demonstrate the Bias of Established Sediment Toxicity Tests, and Develop a Truly Utilitarian Toxicity Test for Sediment-Dwelling Invertebrates

Premise

Conventional sediment toxicity testing, a mainstay of ERA investigation, is predicated on the reality that contaminants that have entered a waterbody do not (predominantly) linger in the water column but rather take up residence at the waterbody's floor, nestled within its recesses (interstitial pores). ERA science is nothing less than enamored with the phenomenon of contaminant sequestration in sediment, and such is evident in the design bias of toxicity testing that routinely proceeds, whether ERA practitioners are cognizant of the bias or not. Without a doubt, sessile forms are most at risk from contaminated sediments for they have no mechanisms to avoid their direct chemical exposures. Reasonably we should expect that sediment toxicity testing would evaluate sessile forms, and certainly more so than they do pelagic forms, the latter threatened through the indirect means of trophic transfer sequences, and also after short-lived contaminant/sediment resuspension events. Nevertheless for ERA, as it relates to contaminated sediments, we find that the field has opted almost exclusively to learn of the fate of free-swimming forms. These include miniature amphipod crustaceans and other invertebrates (e.g., the midge, *Chironomus tentans* in the nymph stage), and immature pelagic vertebrates (e.g., 14-day-old Sheepshead minnow, *Cyprinodon variegatus*) exposed unnaturally to chemical stressors that in the main do not appear in their surroundings. An unbiased newcomer to the field might marvel at the essential design of the universally applied testing arrangement. Site sediment samples transported to the laboratory are agitated to liberate to the column water above, their contaminant holdings, many of which it should be noted, are tenaciously bound to sediment particles. With this separation/extraction step complete, commercially available water column species (vertebrate, invertebrate, or both) such as 1–14-day-old Fathead minnows (*Pimephales promelas*) are then placed into aquaria for prescribed exposure/monitoring periods (e.g., 7 days). Critically, then, we observe that *column water* species are the test organisms of choice in a concerted effort to know if contaminated *sediments* are problematic. In what way, though, does the above-described sequence constitute "*sediment* toxicity" testing? Is there ERA utility in knowing how column water species respond to water that has been unduly enriched with contaminants that were otherwise entrained in the sediment? For argument's sake, would it be just as fair (and also utilitarian) to monitor the responses of sediment-dwelling animal forms (e.g., tubificids) after their substrate had been artificially enriched through the injection of water column contaminants that had somehow been strained from the overlying water? While it is true that in the aftermath of certain events (e.g., boat rudder-generated turbulence), water column species come to be exposed to more contamination than they would ordinarily encounter, resuspension events are not everlasting; in time, the contaminants come to settle again on the waterbody floor. Additionally, not every waterbody is subject to sediment contamination resuspension. Boat rudders don't play havoc with sediments when the water is 30 feet deep; and for a shallow stream (perhaps just 2 or 3 feet deep), only an occasional deer or human trespasser walking across it might come to disturb things. Just as the column water doesn't long remain a murky brown from the disturbed sediment and doesn't blind fish and other species from seeing where they are going, so too are water column species only exposed to otherwise sediment-bound contaminants for relatively brief periods. The upshot of this discussion is that conventional sediment toxicity tests are designed to inform only on the influences of exaggerated and forced contaminant exposures (i.e., exposures that do not actually occur) that are not of interest to us. While pure science might be curious to know how test organisms respond when continuously exposed to contaminants they would not ordinarily encounter, ERA has no such curiosity. In readiness for the study to be described, we should be reminded of certain toxicity testing "ground rules." Sites are not remediated to protect miniature crustaceans, and especially when they may not naturally populate the contaminated site. With this in mind, the astute ERA practitioner will ask: so what, then, if a fair amount of test critters die in the testing? Can one legitimately extrapolate from the test critters' response to those waterbody species that were not tested but that could be the focus of a motion put forward to remediate (e.g., a larger fish)? A second ground rule—it is always the case that toxicity tests report on the responses of organisms that have never previously known contamination. An organism that is all of a sudden thrust into a contaminated milieu is far removed from the sediment-dwelling or water column-dwelling forms that naturally occur at a contaminated waterbody of interest. Are there sediment-dwelling and/or water column-dwelling vertebrates and invertebrates in the waterbody of interest, anyway? Did anyone bother to check, and if the waterbody should be sufficiently populated with the life forms that trained limnologists would expect to be present, why is toxicity testing happening altogether?

This study, as is evident in its title, has two distinct elements. The first is based on the presumption that standardized sediment toxicity testing often wrongly concludes that contaminated sediments are problematic, compromising the health of aquatic species, if not killing them outright. In tandem with evaluating a contaminated waterbody's sediment with the existing toxicity testing arrangement (i.e., employing pelagic species), this first study element calls for an unbiased, alternative, and reality-based testing design to be run. To the extent that the existing testing arrangement is observed to be consistently overstating attendant risks, the likelihood of this occurrence should be calculated. The study's second element is a solicitation for enthused individuals to apply and critically review several described, previously untested methods focused on sediment-dwelling species.

Study Guidelines—Element 1

1. Transport to the laboratory in duplicate, sediment and column water samples of contaminated and control waterbodies, in precisely the same fashion as these are transported in support of conventional sediment toxicity testing (as described in the premise's first paragraph).
2. Process half of the sediment and water samples (i.e., with agitation, etc.) to conduct the conventional toxicity testing with free-swimming forms exposed to contaminant-enriched water. Endeavor to use at least one commercially available vertebrate and one commercially available invertebrate species in the test aquaria.
3. Transfer the other half of the samples—sediment and collected water— to other aquaria, and allow the system to equilibrate for seven days (i.e., with all known sediment particles having settled to the bottom). After the equilibration period, introduce into separate aquaria the same test species as in guideline 2. Ensure that the same number of test specimens are used in the paired testing environments, and that lighting and ambient temperature for these aquaria match those for the aquaria described in guideline 2.

4. Endeavor to conduct all testing (i.e., "conventional" and "equilibrated system/non-disturbed sediment") in the same room to provide for the greatest degree of ambient environment test condition uniformity.
5. Statistically compare all growth and survival endpoints for the paired "conventional" and "non-disturbed sediment" tests.
6. Endeavor to run additional paired tests, modifying only the "non-disturbed sediment" aquaria, and relabeling these the "slightly disturbed sediment" treatments. In a standardized fashion, add to the protocol one or more occasions wherein which sediments are deliberately disturbed, perhaps by inverting the tightly lidded aquaria, or by dragging an object through the sediment for one or two passes. The duration of the standard tests should dictate whether or not planned disturbance can be added to the testing matrix altogether, as well as the number of occasions where there is allowance for incorporating disturbance. Thus, tests of only 48 or 96-hour duration (e.g., respectively, a *Daphnia* spp. LC50 bioassay, and a rainbow trout LC20) do not allow for imposed disturbances. Ten-day *Hyalella azteca* survival tests and 30-day eyed egg-sac early-life stage testing with Salmonids, though, would allow for perhaps two or three disturbances. Pursuant to the testing, statistically compare all growth and survival endpoints for three distinct treatments, as "conventional" (water-only exposures), "non-disturbed sediment," and "slightly disturbed sediment."

Study Outcomes and Applications Thereof

To the extent that the statistical comparisons reveal "conventional" testing to consistently trigger growth and survival endpoints more so than do the "lesser disturbed sediment testing" arrangements, we stand to learn that the former is, in the main, not representative of the true site condition. Here, there will be the opportunity to estimate how frequently in site management work, advanced testing, and dredging and other remedial actions have unnecessarily proceeded. The comparative testing bears the potential to identify a critical tipping point—that degree of physical sediment upset (and thereby, liberation of entrained contamination) needed to trigger toxicity testing failures. The more experimentally introduced resuspensions required to trigger test failures should naturally lead to a refined consideration of the aquatic system dynamics for the waterbodies from which the sediment and column water derive. The need to know if a contaminated waterbody might actually undergo such physical disruption would be recognized. Pursuant to site-specific evaluations, the less likely it would be for sediments to be notably disturbed, and the smaller the size of areas that could be subject to sediment upheaval events, the more obvious it will become that the conventional testing is inappropriate and misleading. Where the biases of conventional testing are exposed, ERA practitioners will seemingly be more receptive to the prospect of developing and applying improved sediment testing that is more realistic and less forced.

Study Guidelines—Element 2

1. Collect, in tandem, sediment samples of contaminated sites and nearby, habitat-matched reference waterbodies. Identify the dominant benthic macroinvertebrate species for paired sites. (Presumably, pollution-tolerant species will have a greater presence in the benthic assemblages of the contaminated sites.)
2. Develop a means to laboratory-rear several of the pollution-tolerant and pollution-intolerant benthic macroinvertebrate species identified at the contaminated sites.
3. Endeavor to conduct all controlled toxicity testing in the same room to provide for the greatest degree of ambient environment test-condition uniformity. Conduct the testing in aquaria as follows:
   - allow field-transported sediments and column water of both clean and contaminated sites to equilibrate for seven days;
   - release fixed numbers of pollution-tolerant and a pollution-intolerant species (perhaps 20 of each) to aquaria for a fixed exposure duration (to be determined; perhaps 14 or 28 days);
   - develop a means for (retrieving and) counting all surviving macroinvertebrates at the end of the exposure period.
4. Statistically compare all growth and survival endpoints for pollution-tolerant and pollution-intolerant species placed into both clean and contaminated sediments.

Study Outcomes and Applications Thereof

Presently, benthic macroinvertebrates living in the wild are not utilized as ERA sediment test species. They are, instead, species that are identified and categorized in comparable sediment samples of matched clean and contaminated sites, and their presence/absence figures into the computation of indices intended to depict the relative health of the benthic communities. Noted differential assemblages at sites—particularly where pollution-tolerant species dominate or constitute the only macroinvertebrates at contaminated sites—traditionally suggest that contaminated waterbodies are biologically degraded. The proposed testing heralds the benthic macroinvertebrate of the field as a test species, with the potential to provide information considerably more utilitarian than that of the current design discussed above (i.e., where commercially available column water species are unnaturally exposed to contaminants artificially driven from their sediment-nested state). The ERA practitioner may come to learn that the absence of pollution-intolerant forms today is a function only of this group having been formerly displaced, and that, if re-introduced, can fare well. The proposed testing may bear the potential to illustrate that ERA practitioners draw errant conclusions from their assemblage-based characterizations of the benthic community. The proposed testing can alert ERA practitioners to the prospect of an overemphasis being placed on benthic macroinverte-brates. The reader is reminded that cleanups do not proceed for the pro-tection of benthic macroinvertebrates, or that an aquatic system's higher trophic levels are unlikely to be compromised where pollution-tolerant forms are dominant.

Study #23

Establish Effectual Sediment Quality Benchmarks

Premise

Procedurally, the evaluation of contaminated sediments in support of ERA should commence with comparisons of site sediment chemical concentrations of appropriate depths with sediment concentrations that have been shown to impinge on the health of the specific invertebrate forms that populate a site's sediment. Utilitarian sediment quality benchmarks, then, should be accurate predictors; identified benchmark exceedances should necessarily translate into direct observations of biological impact in the field and, in the absence of exceedances, sediment infauna and epifauna should appear normal with no signs of stress. It is unfortunate, but, since its inception, ERA has been without reliable and effectual sediment quality benchmarks to initiate assessments at contaminated sites. Why the commonly applied benchmarks fail to meet the task at hand is rather obvious; truthfully they have been inappropriately developed, inappropriately compiled, and inappropriately applied. Brief clarification on each of these drawbacks is in order.

- The hallmark of sediment benchmark development, for better or for worse, involves exposing commercially available (i.e., laboratory cultured) macroinvertebrates, in rather sudden fashion, to contaminated water (as either site/pore water, or artificially adjusted water). This testing system does not in the least mimic the chemical exposures of benthic macroinvertebrates in the field. Natural sediments do not, in the course of just a day, let alone within minutes, become contaminated to the point where they can potentially trigger substantial debilitating effects in macroinvertebrates. A second recurrent feature of standardized sediment benchmark development (as discussed in the previous study), further weakening the science, is the testing with water column/free-swimming species, where these are intended to supply a surrogacy role for sediment-dwelling forms. Is it a fact that water column inverte-brates (e.g., *Daphnia magna*) respond as do sediment-dwelling forms? A third drawback, although often unavoidable, is that a given (water col-umn) test species may likely not be one that inhabits a given ERA site. Do all macroinvertebrates respond similarly? Lest the reader offer that sediment toxicity testing is but a prelude to more engaging sediment effects study, it is not so; many regulators move for sediment remedia-tion based only on the outcome of sediment testing. Should additional work follow sediment testing, such will take the form of comparing ben-thic macroinvertebrate assemblages of site and reference. Is a noted shift towards pollution-tolerant species at a contaminated site indicative of an ecological problem, though?

- Perhaps the most frequently applied sediment benchmark set arranges test and other biological response concentrations in rank order (i.e., list-ing the lowest concentration to trigger a negative response to the high-est concentration to do so). Highlighted are arbitrarily assigned effect thresholds, namely the lower ten percentile and the median. Critically, the values that comprise a set for a given chemical (perhaps 20 or 25) are of a widely diverse nature. Compiled are freshwater and estuarine water data, values from geographically distanced waterbodies of the United States, laboratory and field response concentrations, a broad range of test organisms (from bacteria to nematode to oyster to fish), and, as should be obvious, the responses of highly divergent toxicity testing designs. The compilations beg the questions of (a) the scientific basis allowing for the data arrangement, and (b) the value that lies in screening site sediment chemical concentrations against the compila-tions. (Why would one concerned with PAH contamination in a second-order stream utilize a compiled dataset such as has been described here, when at least half of the values are from bays or still-larger sounds?) Seemingly, more so than constituting a utilitarian effects-screening tool, the compilations might have intended to demonstrate that some form of commonality in biological sensitivity to chemicals exists. For the thinking ERA practitioner, however, the compilations only leave one with the distinct impression that the data was pooled primarily (only?) so as to ensure a sizeable review set. In any event, the benchmarks leave the ecological risk assessor wanting, because of the odd assembly of data points, and certainly when the user again recalls that there is little else to do in sediment assessment for ERA but to screen site concentrations.

- In line with the above, due to the way sediment benchmarks are derived and certainly assembled, their use can only lead to interpretations that field conditions might be stressful to benthic species and possibly, as well, to column water species. Benchmark exceedances cannot and should not be interpreted to mean that, definitively, aquatic impacts are occurring or have already manifested. Too often, however, exceedances *are* applied in just this way. The limitations of even the best of benchmarks must be realized. Hence the present study.

Several interwoven concepts shape the intended study. First, virtually all contaminated aquatic sites in ERA work are long-standing (i.e., several decades old). Consequently, ecological effects of worth, if these were to ever arise, would have already done so by the present day. Second, benchmarks in ERA work are for screening purposes only; the outcome of screening efforts are not risk assessments themselves, but rather indications of whether a site condition will require additional study to verify an unhealthy ecological condition if it exists. Third, that certain sediment invertebrate species might be precluded from populating a waterbody does not necessarily mean that a waterbody is ecologically impaired. With these concepts in mind, this study invites motivated individuals to devise a benchmark set that is effectual such that it can accurately assess when sediment chemical concentrations equate with evidence of the site (sediment) ecology having been compromised.

Study Guidelines

1. No part of the effort to develop the benchmarks being sought should involve toxicity testing of the conventional arrangement, whereby the responses of commercially available organisms placed into laboratory aquaria, are monitored, etc.
2. Recognize that the presence of multiple contaminants in sediment necessarily interferes with (the possibility of) the development of chemical-specific benchmarks. Note additionally that while toxicity testing in the lab for the purposes of deriving sediment benchmarks has the ability to allow for singular chemical exposures, waterbody sediments rarely bear just one chemical.
3. Effectual sediment quality benchmarks, as their name connotes, inform only on the ecological health of the sediment; they do not inform on other components of the aquatic system. Thus, for the subject study, no attention should be given to a contaminated waterbody's fish. Study enthusiasts should recall that no means exist for relating the well-being of a waterbody's fish (as in the species present, population sizes, specimen lengths, etc.) to its benthic macroinvertebrate community. Importantly, too, information does not exist that relates benthic macroinvertebrate assemblages to sediment health. Reasonably, though, where benthic macroinvertebrates (in terms of overall number and in biomass)

are notably reduced relative to the reference site condition, it is fair to assign causation to in-place sediment contamination.

4. Compile a list of contaminated aquatic sites that have yet to be remediated. Review the sediment chemistry data that exists. Segregate sites as those that have only one (or predominantly one) sediment contaminant, and those with multiple sediment contaminants. Record how long it has been since sediments were sampled and chemically analyzed. Eliminate from further consideration sites where substantial sedimentation has acted to bury contaminants such that they can no longer be termed surficial (i.e., occurring in the bioactive zone). Eliminate sites with sediment contamination data that is more than five years old. Sites yet to be remediated but with old chemistry data and/or benthic macroinvertebrate assemblage information can be made to qualify for study pursuant to new data collection efforts.

5. Identify comparable reference waterbodies for the qualifying contaminated sites. These may include upstream components of the contaminated system of interest. Comparable reference waterbodies should necessarily be geographically near to their paired contaminated sites, and certainly within the same region (e.g., the same county).

6. Tabularize benthic macroinvertebrate assemblage information for two types of contaminated sites (and for paired reference sites in all cases), as those with one (predominant) contaminant, and those with multiple contaminants. Endeavor to specifically tabulate the species present, relative species abundance, percentage of species and individuals that are pollution tolerant and intolerant, and a benthic macroinvertebrate biomass measure (e.g., kg of biomass per m^2).

7. Where parity in benthic macroinvertebrate assemblages exists for contaminated sites and paired reference sites, conclude that a chemical's average sediment concentration is safe (and, thereby, insufficiently high to serve as an effect benchmark). Where benthic macroinvertebrates are notably reduced (primarily through a biomass measure) relative to the reference condition, conclude that the average chemical concentration *can* serve as an effect benchmark, providing two criteria are satisfied. First, the disrupted benthic condition must be observed in more than half of the contaminated waterbodies considered. Second, the disrupted condition must be described for at least one waterbody that only has that chemical of concern in its sediment. The second criterion guards against sweeping conclusions that may not follow from the collected data. By way of example, in a situation where four metals exist at excessive levels in surficial sediments, it would be inappropriate to conclude that any one metal is the cause. It would also be inappropriate to set a benchmark for any one, or for all four, of the metals, since metal-specific effect attribution had not been demonstrated.

8. Compare chemical-specific benchmarks from the study to a compilation of published and vetted sediment quality benchmarks commonly accessed for

ERA needs. Identify all instances of noteworthy differences in magnitude between established benchmarks and those generated by the study.

Study Outcomes and Applications Thereof

The study bears the potential to bring forward the information to show that existing benchmarks are ineffective, more often than not suggesting or claiming there is undue stress or demonstrated impact when such does not actually exist. The greater the number of instances where the vetted toxicity thresholds and the study-generated ones are widely divergent, the greater the opportunity there will be for ERA practitioners to reconsider the worth of the former. In this context the simplistic workings of the "Background Approach" to setting sediment benchmarks, one of five approaches considered within the National Status and Trends Program of the National Oceanic and Atmospheric Administration, bears mention. In that approach, chemical concentrations from contaminated (so-called "target") areas are compared against chemical data of pristine areas that are taken to represent the background condition. Where the background values are exceeded "by some specified amount, such are considered unacceptable." The ERA practitioner should find this troubling, along with the following additionally excerpted text:

> In some cases the criteria were set at some value above the background concentration, say, at 125 percent of background or two standard deviations above the mean background concentration. This approach does not involve any determination or estimation of effects.

That practitioners have likely long since forgotten admissions such as the immediately preceding one (if they ever chanced to do the initial background reading to know of these in the first place), is itself inexcusable. These individuals should be fully aware of the origins of the benchmarks they work with, and what exceedances of them signify. Contaminated sediment review cannot hinge on arbitrarily set thresholds, hence this study's essence. Where benthic macroinvertebrates at contaminated sites are supported and occur in comparable manner to that of a reference waterbody, there is no issue to investigate. The study encourages computation of the degree to which existing benchmarks apparently overshoot. Doing so will inform on how frequently a sediment issue is flagged when, in reality, there is no issue to flush out. Benchmark sediment reform can potentially follow from this study. Thus, appropriate benchmark sets to be arranged will need to shed their baggage. As one example, references to fish will need to be weeded out, unless a body of information is created that relates sediment concentrations, specific macroinvertebrate assemblages, or both, to fish population health. (Consider that if testing with flounder [*Paralichthys* spp.] does not proceed in studies leading up to establishing sediment benchmarks, we do not know that a given site's sediment chemistry is challenging to this fish.) Similarly, benchmark sets cannot fuse the findings of diverse waterbodies, and seemingly, as well, the findings of laboratory and field investigations. At a minimum, segregating the data of divergent efforts must occur.

Study #24

Assess the (Potentially Lesser) Nutritional and Ecological Roles Played by Pollution-Tolerant Benthic Macroinvertebrates

Premise

In support of ERA efforts, the evaluation of contaminated waterbodies often extends to characterizing the benthic macroinvertebrate community. Sediment samples of the bioactive zone are collected, after which the invertebrates in the samples are separated from the matrix to be identified (ordinarily) to the level of phylogenetic family or genus. The characterization of the benthic macroinvertebrate community commonly extends to reporting the number of families or genera that are classified as either pollution-tolerant or pollution-intolerant, along with an indication of the relative numbers (or percentages) of specimens that align with these two categories. The benthic community characterization effort also extends to reviewing the species assemblage of a comparable reference (i.e., not contaminated) waterbody. Suitable reference waterbodies (e.g., a stream of the same order as the one that is contaminated) should necessarily be located geographically near to their contaminated counterparts (e.g., in the same county), and, depending on the circumstance, may be a component of the same aquatic system as the contaminated one. Thus, an upgradient (non-contaminated) portion of a stream, or a somewhat-removed river reach, may be quite adequate. In lentic systems (e.g., lakes) with "hot-spot" contamination, areas distant from the zone of contamination may make for adequate, if not preferred, reference locations.

Where contaminated aquatic sites display benthic macroinvertebrate assemblages with distinctly greater proportions of pollution-tolerant forms than expected, often the interpretation is that the site is stressed with ecological function at a subpar state. In these instances it is suggested that fish are being shortchanged in terms of diet, leading in turn to lesser fish productivity. Are pollution-tolerant benthic macroinverte-brates, though, less nutritious than pollution-intolerant ones for fish that feed at this trophic level? Do the former supply less energy than the latter, and, if so, do we find fish size, fish health, fish population size, fish spe-cies diversity, or any other metric or correlate of ecosystem function to be compromised because of the contamination-based shifted diet? This study is seeking the answers to these and other related questions. Potentially, the merits of investing energies in characterizing the composition of benthic macroinvertebrate assemblages is up for grabs; there may be no purpose at all to computing relative biodiversity indices (e.g., Shannon-Wiener) because other than for the benthic macroinvertebrates themselves, aquatic sites are as functional as their reference site counterparts. This study nec-essarily has field and laboratory components that should be conducted in tandem.

Study Guidelines—Field

1. A complete study would not only involve data collection of multiple waterbodies of a certain type, but would also include data collection of varied waterbodies (i.e., lakes, streams, rivers). As with earlier described studies in this book, this study creates great opportunities for multiple-party participation through which a more comprehensive analysis can ultimately emerge.
2. Several contaminated waterbodies yet to submit to remedial activi-ties should be sought out. Certain obvious advantages would accrue to selecting waterbodies that are part and parcel of Superfund investiga-tions and the like. At a minimum, waterbodies with such a regulatory background likely have highly available, utilitarian information on chemical history and benthic macroinvertebrate community structure. A cautionary note: if benthic macroinvertebrate assemblage informa-tion for any waterbody is not up-to-date, study complications may arise when using it.
3. Waterbodies that are subject to periodic fish stocking or that have other than catch-and-release fishing programs, while still serviceable in that they can supply certain useful data (see guidelines 5 and 6), will, quite obviously, not be able to support guideline 8.
4. Although not absolutely necessary, an effort should be made to sample the sediments of comparable non-contaminated waterbodies serving as reference locations. Where this is done, samples from these locations should be of the same depth, and be collected at the same time of the season, and via the same method (e.g., kick net). Samples should have

all non-macroinvertebrate materials removed, followed by rinsing and drying the macroinvertebrates in preparation for an energy assessment with a bomb calorimeter. Efforts should also be made to compose samples that aggregate specimens of singular macroinvertebrate species identifiable to the family or genus level.

5. Benthic macroinvertebrate samples of the two types described in the preceding guideline (i.e., mixed species; individual species) should be assessed for protein content and carbon content, both per unit mass.

6. In a structured manner, macroinvertebrate biomass should be measured in representative sediment samples of uniform size and depth from areas that have (a) little or no pollution-tolerant forms, (b) moderate pollution-tolerant species representation, and (c) dominant pollution-tolerant species representation.

7. The information of the earlier steps should be tabularized to assist with identifying possible linkages of macroinvertebrate nutrition (protein content), energy stores, biomass, and macroinvertebrate category (as pollution-tolerant or pollution-intolerant).

8. The fish community should be rigorously described and evaluated for numbers of fish, fish size/weight, and age (by species) to support a comparison with waterbodies that differ in their benthic assemblages (i.e., as per the categories identified in guideline 6). A concerted effort should be made to identify correlations or regressions for the macroinvertebrate measures (guideline 7) and fish statistics (this guideline).

9. This study is directed at integrating the influences of three primary phenomena, namely:
   - the co-occurrence of sediment contamination and pollution-tolerant macroinvertebrates;
   - the possible lesser nutritional and/or energy value of pollution-tolerant macroinvertebrates over pollution-intolerant forms; and
   - the extent to which a lesser fish compartment (in terms of size and/or quality) is attributable to a lesser-quality macroinvertebrate diet.

   A fourth phenomenon to be incorporated into the analysis is chronology, i.e., specifically considering how long studied sites have been contaminated, and how long fish have been feeding (we assume) on what may/may not be found to be a less nutritious diet than was previously available.

Study Outcomes and Applications Thereof

While it is known that contaminant influxes to waterbodies can displace pollution-intolerant benthic macroinvertebrates, it is not known if such community shifts directly impact invertivorous fish. This study should furnish the information to either support or refute the often-stated, but as yet unproven, thesis that fish and aquatic ecosystems, overall, are degraded where pollution-tolerant macroinvertebrates have been introduced and possibly have a dominant presence atop or within the sediment.

Where invertivorous fish of contaminated waterbodies are not found to be impacted, the presence and role of pollution-tolerant benthic macro-invertebrates in the sediment will be understood to be unimportant. Independent of what a nutritional analysis might show, the phenomenon would be evident. Additionally with this outcome, the historical over-emphasis placed on sediment toxicity testing in support of ERA work will be seen to have been quite unnecessary. The veracity of the argument put forth—that aquatic ecosystem adjustments to stressors occurring over the decades are what explain ample fish populations—will need to be rec-ognized. The reality is that all sites have been contaminated for multiple decades by the time they stand for review.

Where a waterbody's fish component is observed to be lacking, the reporting will need to address the causal factor, as either (a) the system contaminants present, (b) the shifted diet, or (c) some combination of these two. Presumably, information gleaned from the study's laboratory compo-nent can help to resolve the question. Where it is only the benthic commu-nity (and not the fish) that has been impinged upon, ERA practitioners and regulators will be forced to come to grips with the import of this change. The following questions that may come to bear should be discussed in the peer-reviewed literature and should necessarily influence ERA prac-tice and guidance, as well as remedial program guidance: do we clean up waterbodies only so that they can be returned to a condition whereby they (supposedly again) support pollution-intolerant macroinvertebrates? How could an affirmative response to this question be defensible given that benthic macroinvertebrate species are not endangered on any areal scale? Is there a pressing need to restore pollution-intolerant forms when waterbodies without their fullest representation function adequately in an ecological sense nevertheless? Do pollution-tolerant macroinvertebrates, pound-for-pound, provide more biomass than do pollution-intolerants, and, if so, does this phenomenon compensate for any lesser nutritional value the former may offer? In that case where a given waterbody's only ecosystem "impact" is a lesser presence of pollution-intolerant benthic macroinvertebrates—something that can only be known if sediments are analyzed by hand—does a need for cleanup exist? Why should cleanups proceed for differences that are essentially invisible, and particularly when all other ecosystem elements appear to be fully supported?

A potential outgrowth of the study may be the recognition that lesser biodiversity indices *for the benthic macroinvertebrate community*, real as these may be, are misleading indicators; they do not inform on the functional health of the larger aquatic ecosystem. Where such is amply established, a fuller argument in support of the ERA reform to dispense entirely with evaluations of the community should be developed and marketed, as follows. Although benthic forms provide the food base for higher trophic levels, sites are not remediated when benthic macro-invertebrates are found to be stressed. Such sites are certainly not remediated if, despite the displacement of pollution-intolerant forms and biodiversity indices found to be askew, waterbodies nevertheless support the larger biota they should. The situation is no different for terrestrial sites. Insect communities are never assessed, although, potentially, insect species assemblages in contaminated soils are nota-bly different from those at nearby reference locations. Consequently, sites are never remediated because of any contamination-posed influ-ences to insects, should these exist.

Study Guidelines—Laboratory

1. The simple objective is to monitor growth and survival of invertivorous freshwater fish that are either directly fed or allowed to feed on either pollution-tolerant or pollution-intolerant macroinvertebrate species that ordinarily populate sediments. Contaminants need not necessarily be a component of the testing matrix.
2. Several fish species should be used. Aquaria should house only one fish species. A defensible number of fish (of a given species) should be set in each aquarium.
3. Macroinvertebrates to be made available can be harvested from the field from waterbodies that are free of contamination, or purchased from a commercial supplier.
4. In the main, juvenile or immature fish are those to be tested/monitored. For matched aquaria (in terms of size, water volume, ambient tempera-ture, lighting, duration of exposure), feeding regimes may either be standardized or *ad libitum*. A defensible basis for the feeding regimes used (which may vary by species, it is recognized) should be supplied. Feeding might best be facilitated with pre-prepared pelleted formula-tions of pollution-tolerant and pollution-intolerant species.
5. For growing fish, weights and lengths, and overall appearance and health (and, of course, survival) should be recorded periodically. The same measurements should be recorded where the fish used are either nearing or at full size.

6. Enthused parties may wish to explore advancing the project to the outdoors, utilizing raceways. The ingredients of other pelleted food formulations that may be used in addition to the macroinvertebrates should be cautiously reviewed. Often, PCBs are an unavoidable ingredient. Note: with the only study variable being the category of macroinvertebrate (in terms of pollution tolerance), the influence of potentially confounding influences (such as PCBs in commercially available food pellets) should be minimal.

7. Presumably, where fish lengths and weights will be found to differ for paired groups, it will be that the smaller/lesser physical measures will align with fish that consume pollution-tolerant macroinvertebrates. In anticipation of such a finding, develop, *a priori*, defensible standards for the degree(s) to which length and weight need to be reduced for the changes to be biologically significant.

Study Outcomes and Applications Thereof

As with the data to be generated for the field component of this (overall) study, statistically supported biologically significant differences between paired fish groups will decide the matter of whether merits accrue to characterizing the benthic macroinvertebrate community. Potentially, fish health is short-changed where pollution-tolerant macroinvertebrates dominate waters (and where we will assume that such a condition derives from site waters and sediments being contaminated; something that may not be easily elucidated). Should fish metrics (for growth and population size as indicators of health) prove to be about the same whether the predominant food source is of pollution-tolerant or pollution-intolerant form, the routine (and at times excessive) attention given to benthic macroinvertebrate assemblages will be challenged at its root. We would learn that the productivity of higher aquatic species is unaffected by assemblage shifts, real as they might be. Among other things, such a circumstance would invite interested parties to explain why the shifts do not translate into reduced fish communities, and particularly if it should be that pollution-tolerant forms are less nutritious than pollution-intolerant ones. Should metrics indicate that fish that predominantly or only feed on pollution-tolerant species are less robust than those that predominantly consume pollution-intolerant species, the "so-what?" question must be dealt with. As examples: does it matter that smaller fish are less healthy, if indeed they should be? Does reduced fish size translate into any other site-based ecological effects that need to be corrected?

To get at this study's full worth, there are still two more considerations. Perchance, study data will reveal a finding that runs contrary to what practitioners anticipate, namely that fish mostly or exclusively consuming pollution-tolerant macroinvertebrates are larger and/or more robust than fish that consume pollution-intolerant macroinvertebrates. Perchance, pollution-triggered benthic macroinvertebrate community/assemblage shifts are good things. Researchers must be willing to acknowledge the arrival of such outcomes, counterintuitive as they may be. Researchers will then have the task of supplying an ecotoxicity-based rationale for the outcome, and they will, of course, be free to expound on what the unexpected outcome might purport vis-à-vis the role of, or need for, aquatic assessment as part of ERA. Finally, researchers must acknowledge that aquatic site cleanups do not proceed because assemblage shifts are identified. Perhaps this follows from benthic macroinvertebrates not being readily observable. Importantly, real as any community shifts may be, it is only when ecological manifestations are observed in fish that one can legitimately say that an issue to be dealt with exists.

Study #25

Revisit FETAX (aka FETAX without Embryo Dejellying)

Premise

Toxicity tests and bioassays in support of ERA are intended to provide indications that contaminated site conditions are health-offsetting to biota. Well-meaning as the tests and bioassays may be, the testing environment cannot help but constitute a departure from the actual field condition. As examples of this, a tightly controlled ambient laboratory setting is simply not the outdoors, and animals in the wild do not find themselves areally confined to cages and other artificial enclosures. While tests and bioassays intend to mimic natural, albeit contaminated, settings as best they can en route to discovering their potential to elicit health effects, they should not be placing test species at unfair advantage. There is no need to know if/that health effects are triggered where a test design involves the manipulation of biological tissue at the hand of man's deliberate intervention. Further, manipulating living tissue to facilitate or elicit toxic responses does not inform in a meaningful way. With this in mind, it is never too late to subject to scientific scrutiny what may be a vetted method, and certainly if there exists just cause for questioning a test method's workings and the interpretations that follow from its application.

The Frog Embryo Teratogenesis Assay-Xenopus (FETAX), first developed in 1983 and ASTM International-approved, is one of the longer-running toxicity testing tools available to researchers seeking to understand the influences of contaminants on aquatic organisms. Briefly, young embryos of the easily reared and hardy South African clawed frog (*Xenopus laevis*), are placed, 25 at a time, into small Petri dishes containing 10 mL of a test solution, and monitored over 96 hours for evidence of abnormal morphogenesis. As with any toxicity test, FETAX has its share of unavoidable features that detract from its power to inform. To the extent that anurans are a concern at contaminated sites, *Xenopus* spp. is unlike the forms found in the United States and the Western world. *Xenopus* is not amphibious, but rather wholly aquatic. *Xenopus* breeds throughout the year, while other anurans are seasonal breeders only. Putting these differences aside (if we can), and also the method's reliance on artificially inducing the release of sperm and egg via needle injections of human chorionic gonadotropin to the dorsal lymph sac, there is one particular procedural step that cannot be ignored—one that gives rise to this proposed study. Prior to being placed in the test solution to begin the assay, midblastula to early gastrula embryos are dejellied through being placed into, and gently swirled within, an 8.1-pH-adjusted, 2% w/v solution of L-cysteine. The purpose of this maneuver is obvious; there is a keen interest in knowing if embryonic development can be offset by a test chemical, or by the one or more chemicals that make a waterbody a contaminated site of interest. Further, and by all appearances, the thirst to acquire that knowledge somehow gives license for a deliberate tampering with nature, namely the removal of a naturally present protective outer embryo coating. Though not intended as such, FETAX, as it has been applied for some 30 years, really only answers a non-ERA-relevant "what if" question: what happens to embryonic development if newly formed frog embryos have their protective outer cell layers removed by being placed into a bath that could, as per the assay procedures, be as much as 39 times more alkaline (and caustic) as the water in which they formed? Truly speaking, ERA is not interested in the answer to such a theoretical question. ERA would only be interested if, as part of natural development, young embryos found themselves transferred for several minutes from their natural home to some other waterbody with water quality characteristics that dissolve the jelly coat, after which the embryos are returned to their "home." ERA would be interested in the assay's outcomes if *naturally*, midblastula to early gastrula embryos somehow shed their jelly coats. Do they, though? This proposed study is seeking a structured comparative review of FETAX outcomes, pitting method-prepared embryos (i.e., with jelly coats removed) against fully intact ones (i.e., with the jelly coats untouched). To the extent that intact embryos display a relatively higher threshold for teratogenic insult—an anticipated finding?—what would seem to constitute several prudent secondary/follow-on studies are described.

Study Guidelines

1. In the main, the study is asking that paired FETAX testing occur within a multiplicity of experimental and natural circumstances. Each paired test would expose newly formed embryos, both with their jelly coats

and without them, to identical test/chemical treatments. Test solutions should be prepared exactly as described in the ASTM-International standard. Seemingly, the greater ERA interest in paired testing outcomes would be where the water of a contaminated waterbody (e.g., a lake) is the "test solution." Considering the assay's miniscule Petri dish test volumes, it would be prudent to run several sample replicates from waterbodies whose remedial fate is the focus of discussion. Seemingly, waterbody testing should require more replicates than are needed when researching test solutions. Whereas range-finding responses are the interest when a test chemical is being reviewed, likely necessitating dilutions of stock solutions, waterbody test samples should never be diluted. Dilutions of waterbody samples would amount to reconfigurations of the natural exposure environment, and, importantly, aquatic receptors are continuously enveloped by their aqueous medium "as is." FETAX applied to ascertain if a waterbody interferes with morphogenesis needs to assess potential endpoint development of the natural-state water.

2. In all cases, the FETAX method's Teratogenic Index (TI) should serve as the assessment tool for embryo wellness.

3. Where test chemicals are being studied (as in traditional range-finding testing arrangements), the interest is in identifying chemical-specific thresholds-for-effect for the two embryo types. More specifically, researchers should endeavor to identify that (anticipated) added degree of resilience to developmental/teratogenic insult that the jelly coat naturally affords. To the extent that added developmental effects resilience (or "lost developmental-effects resilience," from the vantage point of traditionally prepared embryos) accrues to non-tampered-with embryos, a trends analysis effort should be assembled. For different chemical groups (e.g., metals, organic solvents), the inquisitive researcher should determine notable and/or predictable levels of resilience-to-effect, if such are apparent.

4. To the extent that added developmental effects resilience accrues to intact/non-dejellied embryos, one or more TI correction factors should be established. In practical terms, these would indicate how much higher than presently thought water column concentrations of test chemicals would need to be for developmental effects to occur.

Follow-On Study Guidelines

1. Given this book's ERA focus, an essential task would involve revisiting past ERAs that included FETAX work, i.e., where the "test solution" was lake, river, or stream water drawn from a contaminated site. The effort should strive to quantify the number of instances where FETAX outcomes likely overstated the potential for aquatic impacts, and the

percentage of FETAX applications that (a) wrongly and unnecessarily identified an ecological risk or impact situation, and/or (b) advocated for a remedial action.

2. A concerted effort should be made even at this late date (i.e., more than 30 years since FETAX's introduction, and at a time when the method has fallen into disuse) to establish direct linkages of unacceptable TIs observed in intact embryos and abnormal morphogenesis appearing in other aquatic forms (e.g., fish). This task would involve vast step-wise laboratory (aquaria) testing to establish that test solution concentrations that yield failing TIs also impact (developmentally or otherwise) fish and other aquatic test species. In all instances, it is imperative that chemical concentrations of test solutions are those that could realistically occur within a contaminated waterbody. The reference to the step-wise laboratory (aquaria) testing (i.e., intact frog embryos first being used, and then subjecting fish to the same test concentrations) being vast recalls the need for testing chemical mixtures.

3. Given the numerous ways in which *Xenopus* spp. is unlike anurans in the United States and the Western world (enumerated in the study premise), an effort should be made to bring online a working teratogenesis assay for an amphibious anuran.

4. All water quality criteria based on the outcomes of historically applied FETAX (i.e., where embryos have been deliberately dejellied) should be retested, painstaking as such an undertaking might be. Perchance, the application of numerical adjustments as discussed in "initial" guideline 4 would obviate the need for a significant degree of retesting.

Study Outcomes and Applications Thereof

Potentially, the information to be brought forward from the related studies will demonstrate that the FETAX assay, as designed, is unnecessarily stringent where it needn't be. Standard FETAX application that provides for skewed and overly protective aquatic assessment should, at a minimum, promote a much-needed dialogue on the purpose and design of toxicity testing in ERA. Beyond amending FETAX where the method has been demonstrated in one or more ways to be unnecessarily identifying unhealthful aquatic conditions, a cautious review of other established toxicity tests and bioassays should proceed. The review's purpose should, of course, not be to discredit the past work of dedicated scientists, but rather to uncover a design element that detracts from method utility in an otherwise intended ERA-support context.

Section III

Desktop Studies

Photo by Christina Graber.

Study #26

Determine the Ecological "Ingredients" that Constitute a *Bona Fide* Site for Study

Premise

Not uncommonly, ERAs include information that communicates to the reader an awareness that one or more analysis components were unnecessary. This information is classically provided towards the rear of ERAs, as in the summary or conclusions sections. As examples, ERAs will note that (a) certain considered (i.e., food chain-modeled) receptors, because of their excessive home ranges, would not be expected to frequent the site, or (b) the outcome of a plant benchmark screening exercise need not be considered because the site is heavily vegetated. While we can commiserate with those who legitimately feel they've had their time wasted after having read about HQ calculations and other evaluations that needlessly proceeded, the recurrent phenomenon can serve as a springboard to a highly constructive exercise. Though contaminant releases may be undeniable, not every site that stands to be listed or that has already been listed within a given regulatory agency's remedial program truly needs to submit to some form of ecological health assessment or review. How highly utilitarian it would be to have a scientifically supported, assembled schema by which sites to potentially submit to ERA could be gauged for their worthiness in this regard.

This study to be undertaken empowers the interested scientist to develop and compile criteria in support of an initiative thus far unexplored. Through a rational and deliberate approach, the idea is to secure threshold criteria to be met, thereby allowing a site to embark on its meander through an agency's remedial program; in short, to determine what qualifies a site to be a site. The interested scientist should consider the following two points: 1. Presently, every site that bears contamination (to include one measuring just 0.01 acres, i.e., 20 feet by 20 feet) submits to an ERA of some sort, and realistically it cannot be that every site needs one.

2. Just as we apply screens *within* ERAs (e.g., to hone down a contaminants of concern (COC) list or a receptors-of-concern list), there is every reason to employ a screen *before* an ERA is conducted, to determine whether or not an ERA should proceed altogether.

Study Guidelines

1. The study bears the indispensible requirement that enthused parties shed the bias that, regardless of circumstance, every site that has been the recipient of one or more chemical release events must—for the purposes of learning of the possible ecological toll taken—submit to an ERA. With indisputable evidence to support site contamination (frequently) *not* equaling ecological impact, such thinking is archaic, to say the least. (It bears noting here, then, that several U.S. states [e.g., Pennsylvania, Washington] articulate a minimum site-size threshold for sanctioning ERAs.) Further, where it is summarily assumed that contamination cannot help but impact the ecology of contaminated sites, those engaging in this project would not be looking for the discrimination between or among sites that is described in the study premise.
2. Seemingly, there are multiple ways to develop the criteria, as well as multiple formats for applying these. The developed product could take the form of a checklist, or something more sophisticated, such as a flow chart, dichotomous key, or a computer program or model. Realistically, there is no one "right" or "best" answer (i.e., a means of arranging and populating a developed threshold criteria schema).
3. While certainly not a scripted study to be taken on, a recommendation is to first compile a list of essential ecosystem components for the contaminated site as it, at least cursorily, is known. It may be obvious, for example, that, due to a minimal site size, certain ecosystem features are comparatively unimportant because they supply a lesser contributory role. Procedurally, the list may be pared down through the elimination of site features that are relatively difficult to collect, and that may not be commonly applicable to sites. See next guideline.
4. Enthused scientists should not shy away from including site elements, the measurements of which are not easy to obtain. Additionally, though it may become apparent that the ERA community would be most unlikely to ever adopt a schema that necessitates the collection of what might be termed esoteric information types, the suggested study is asking that such considerations be ignored. Those locking onto the study should give thought to developing two schemas, one that reflects what many ERA practitioners could only reasonably embrace in the way of data collection needs, and one that could likely call for the collection of unconventional and hard-to-obtain data.

5. To the extent that features to be considered in developing threshold criteria for sites warranting an ERA vary considerably by specific site type, and to the point where a dichotomous key would not be helpful, consideration should again be given to developing multiple schemas. Sites with extreme climates, for example, may not be adaptable to what might otherwise turn out to be a genericized guide or schema.

6. It is recognized that a proper end product is not necessarily one that incorporates an exhaustive array of criteria. Finished-product schema(s) may likely reflect criteria that were originally considered, but that were subsequently found to not play into determinations of the need to assess or not assess ecological risks. Finished-product schema would be their most complete if they were transparent, in the sense that they explained why certain data types were eliminated from a decision matrix.

7. It would seem that a short list of indispensible criteria for a functional schema would include, anticipated or actual species present, anticipated or actual population sizes, a grounded expectation of the species diversity a site should display, knowledge of how long a site has been contaminated, and knowledge of how long it should take for a critical ecosystem component to become compromised.

8. This study, like the one presented before it, lends itself to multiple interests taking on the challenge to arrive at the practical sought-after schema. There would be every reason for a party to work at developing a schema while knowing that another party had already produced one, and even where it had been published in the peer-reviewed literature. There would be great value of second, third, or fourth interests coming forward with schemas, and particularly where these might be notably divergent from one another—either arriving at the same point through alternative means, or arriving at decidedly different points. Parties taking to this study's technical assignment/challenge would likely intend on having their findings published. Owing to the novelty of developing a defensible means for segregating sites that should and should not be assessed, parties realistically need not worry about peer-reviewed journals being less than enthusiastic in the event others have already published in this vein. If nothing less, a critical-review article consolidating the findings of multiple attempts to develop a schema could well find a home in a respectable peer-reviewed publication.

Study Outcomes and Applications Thereof

Study outcomes could likely discover that there are considerably more cases of Superfund-type sites that do not warrant (even) a simplistic ERA than most practitioners would imagine. Where such is the overall take-home message, a concerted effort should be made to secure the approximate frequency with which historically, sites have unnecessarily been put through the ERA process. This follow-on task would admittedly be labor-intensive in that it would require obtaining a working familiarity with sites not unlike that expended for the schema development effort. Should the implication be that, from this point forward, great proportions of sites will be anticipated to not be in need of ERA, another responsible and prudent follow-on task would present itself. A stratagem should be developed for having the newly gained information infiltrate the ERA guidance of the relevant regulatory agencies, with the goal of only having ERA efforts proceed where ecological impacts of import could realistically arise.

Study #27

Determine a Site-Age Statute of Limitations for Submitting to ERA

Premise

Characteristically, the period between a first contaminant release event occurring—seeding a site to be assessed for ecological concerns sometime in the future—and a discovered and prioritized contaminated site submitting to its first ERA investigation element, encompasses several decades. During that protracted period, myriad environmental processes affecting a site's abiotic and biotic site components will have been actualized. Chemicals may have degraded to less toxic daughter products (although the possibility of more toxic daughter products arising should not be summarily discounted), chemical concentrations may have decreased, and chemicals may have become tightly bound to substrates such that they are minimally bioavailable. In the aftermath of tens of generations occurring over the decades, the biological species that concern us and that ERAs evaluate in some way have had recognized opportunities to adjust to the contaminated site condition. To be fair, there are grounds to suggest that there may be no need for ecological assessment work (at any scale of endeavor) for aged program-listed sites.

This study calls for a detailed exploration into the science of the development of chemical-posed health effects as these necessarily occur in the field (assuming they do), the ability of sites to sustain their chemical stressor impositions on biota, and the pace at which chemical-posed effects wane. The latter two aspects would need to consider the constancy or minimization of a site's chemical footprint, and the biological/genetic capabilities to outbreed stressor effects.

Study Guidelines

1. The review or conduct of laboratory dosing experiments should play no part in this study, for the forced chemical exposures, artificial environments, and general design elements of such efforts bear no relation to the field condition. The utility of laboratory-based experimental designs particularly falls away in the realm of (an absence of) successive generations of exposure, which constitutes the essence of this study.

2. In the main, approaches to study are of two types: bottom-up and top-down. With the former, the effort is made to apply existing science to ascertain how many years it would take for a contaminated site to first elicit health and population impacts. Implicit in this exercise is the understanding that contaminated sites do not arise "overnight"; rather, repeated chemical releases occurring over a period of years is what gives rise to sites. Researchers then, should explore the phenomenon of site biota being conditioned to contaminants, because these only accumulate gradually. Researchers might prudently investigate the possibility of site biota capability to adjust to the gradual infusion of contaminant loads, and to the point that health effects fail to be elicited. The latter (top-down) approach necessarily relates to aged contaminated sites that either do or do not presently demonstrate health and population impacts. Where effects are still occurring, the objective is to generate reasonable estimates (projections) for these arising with lesser and, at some point, inconsequential frequencies. A mainstay case is anticipated for the top-down investigation, namely that it will be difficult to identify sites that are presently eliciting health and population impacts. For contaminated sites that do not display health or population effects, a worthwhile investigation component seeks to determine if these sites did, at one time, display effects, or if they did not do so.

3. It should be clear that effects of interest, for all intents and purposes, are reproductive, recognizing the myriad ways in which this key biological function can be offset (e.g., a lessened intrinsic rate of population increase, a developed irregularity of the ovulatory cycle). Researchers are reminded that accumulations of contaminants in tissues (e.g., the concentration of cadmium in the femur) are not health effects. The exception to this would be that case where, for a given contaminant, the concentration within a certain somatic compartment that causes a serious health condition is known.

4. Study efforts need to embrace both terrestrial and aquatic sites.

5. Researchers should be cognizant of the distinct possibility that contaminated sites demonstrating population and other impacts may be difficult, if not impossible, to happen upon. Where such is the case, the study need not be aborted, but adjusted to nevertheless make advances in the topical area (hence guideline 7).

6. Researchers should be cognizant that existing contaminated sites that either demonstrate ecological stresses or impacts or that appear to be free of same may be so areally small as to limit the gains being sought. To the extent that a great percentage of contaminated sites are areally small, the study's statute of limitations interest will necessarily morph, but not altogether disappear. A successful study end-product would be a thesis integrating site-size considerations, and considerations for investigating ecological effects at sites where considerable time has lapsed.

7. Creating a contaminated site: electing to chemically adulterate sites through deliberate chemical addition would constitute a *bona fide* and perhaps recommended approach, for the concept of experimental forests is well established. Researchers need to be wary of the unavoidable downside of this study option however, namely that appreciable time must be allocated to lapse before investigation inroads can be made in ascertaining the pace at which triggered health effects dampen out. (Truthfully as well, time is needed to allow for effects of concern to first develop.) Creating contaminated sites bears its own brief list of guidelines.

   • Sites need to be dosed in a reasonable manner; primarily this means that it would be unfair to introduce chemical concentrations of extreme magnitude through a singular application or through just a very few of these. Such site dosings would be equivalent to the commonplace one-time bolus injections administered to laboratory animals in drug trials. Importantly, such aggressive site dosing would more than reveal the intent of researchers to accelerate site dynamics so that data on the dampening of effects can be collected that much sooner. It would make most clear that the dosed ecosystem would not be replicating the field condition of sites that ordinarily submit to ERAs (presently). At a minimum, the biological/biochemical coping mechanisms that species call upon when toxic threats are apparent would be unfairly overtaxed. Importantly, too, because the chemical additions would probably not be administered appropriately, not all ecosystem compartments (e.g., decomposers) would, at least initially, be adjusting to the changed site dynamics.

   • As part of the overall study analysis, care should be taken to include a distinct and complex, albeit generally infrequent, site arrangement, namely one where exposure-point concentrations are still increasing because a contaminant release process is ongoing.

   • To avoid potential confounding factors, a concerted effort should be made to ensure that selected land parcels and aquatic habitats to be experimentally dosed have had no known prior contaminant releases.

   • Without a doubt, there are work-arounds for those enthused with the option of creating contaminated sites for the purposes of achieving gains regarding the overall study goal. Creatively, terrestrial and aquatic systems can be dosed such that all depths of soils and sediments bear contaminants, and with depth profiles that actualize real-world conditions (e.g., decreasing concentration with depth).

8. Given the unique scope of this study and the myriad ways in which the subject matter could be explored, the study is ripe for having multiple independent research interests putting their best feet forward.

Study Outcomes and Applications Thereof

To the extent that multiple interests engage with the study topic, a consolidation of the independent findings of these should be produced. A thorough review would identify "site-age statutes of limitation" relevant to terrestrial and aquatic sites to otherwise summarily submit to ERAs, with the reported figures (expressed in years or decades) sensitive to the various categories of sites that exist. Reasonably, the pace of chemical breakdown and animal adaptation (among other processes at play) vary by ecotype (e.g., savannah, grassland) and with the chemical footprint (e.g., predominantly organics, predominantly inorganics, only a few chemicals present, many chemicals present). Potentially, the information to be brought forward could greatly impact the ERA field; for the general case or perhaps in all instances, the information may decisively reveal that most or all sites are too old to legitimize a need for ERA. Such information could notably trigger ERA reform whereby guidance would be set forth, instructing on those instances and site conditions where ERA is not warranted.

Study #28

Develop a Process for Site Physical Stressor Assessment in ERA

Premise

As environmental stressors go, clearly and understandably, due attention has been given to chemical footprints at sites. Having toxicity assessment as a fixed step in the ERA process secures that chemicals released to the environment garner their necessary attention. Notably, however, and going back as far as the earliest of ERA guidance documents, there is an expressed intent to evaluate all of a site's environmental stressors for their contributing capacities to impinge on site biota health. For one thing, where applicable, the influences of radionuclide releases at sites have been included in ERA work. As it turns out, in the overwhelming number of cases, radionuclide contamination is not an issue at sites, and determining if there should be such an issue is rather easily decided. Sites can be walked with portable equipment to arrive at simple radioactivity presence (above background)/absence determinations. Following from the consideration-of-all-potential-site-stressors "mandate," physical site stressors must also be reviewed, for they also have the capacity to impact site ecology. Seasoned ERA practitioners should reflect on their past site-specific engagements to arrive at a loose figure for the number of sites where they have seen physical site stressors discussed where it was appropriate to do so. They should reflect, too, on the number of sites where they have witnessed physical site stress integrated into applied assessment schemes. Chances are that attention to physical site stress, if it was discussed anywhere, was confined to brief mention in a singular paragraph of a site's ecological setting description. It is curious that there has been such inattention to this site component; while it admittedly may not always be a factor at play, it can unquestionably combine with chemical stress to exacerbate ecological effects in some way. Physical site alteration at the hands of man can be long-standing, defined here perhaps as being five or more decades old.

While global climate change has also been ramping up in recent decades, this (thermal) form of stress has been receiving great attention, while physical site alteration has not. Thus, chemical effects testing with animals exposed to different thermal regimes regularly occurs, and ecological outcomes to receptors in the wild are being projected as environments continue to grow warmer. It is curious, then, to say the least, that formal means for incorporating the effects of physical interferences into present-day chemical-based ERA has yet to be pursued. While the ultimate in this spoken-of incorporation would involve assessment schemes that integrate the influences of chemical and physical stressors, we should duly note that, taken as a lone stressor, physical site alteration and habitat impact arising from it have yet to submit to a formal assessment scheme.

This study is reaching out to enthusiastic scientists to develop a mission that the regulatory bodies have either overlooked or perhaps deliberately avoided. In the latter case, it may be that the sheer complexity of the topical area, to include the myriad forms that site physical alteration can assume, has led to inaction. Site physical stress, though, potentially offers an attractive feature, namely that it can and will often remain unchanged for great amounts of time, a situation quite unlike that of chemical footprints in environmental media, where reduced concentrations and morphed chemical states are commonplace. In contrast to chemically contaminated "moving target" sites, where data just a few years old may no longer be relevant, physically impacted sites boast a constancy of the stresses they may be imposing. Consider that the legacy effects of years of tank and other military vehicular use, to include severely compacted soils and a highly irregular site terrain (of mounds and great depressions), will certainly not, of their own, revert to an undisturbed and more uniform land surface/topographical condition. Admittedly, this study is the most ill-defined of this book's studies compilation, and this follows as much from the variable nature of physical site impairments as it does from no one having previously lifted a finger to attack the (seeming) problem area. You, the enthusiastic scientist, can be the first to introduce inroads to ERA science where they are so needed.

Primary Approach Study Guidelines

1. For both terrestrial and aquatic sites, compile a broad list of man-induced and man-emplaced physical impairments.
2. For each of the impairments, assign one or more ecological impacts that could reasonably be expected to ensue. In a developing table or matrix,

supply a column with entries that describe supposed mechanisms for the expected ecological effect(s).

3. Supply a next column that lists specific tell-tale data types (e.g., animal specimens, tissue measures, testing outcomes, computed ratios, etc.) that should identify ecological effects that stem from imposed physical site features, if such should be present.

4. Locate a series of test sites, both terrestrial and aquatic, some of which only bear (released) chemical footprints, some that bear only physical stressors, and some that bear both types of stressors. Document, as can best be done, how long chemical and physical stressors have been in place. Identify sites to serve as viable reference (non-stressor) locations. These should be habitat-matched, understanding, of course, that the physical stressors at the test sites have, to some extent, necessarily compromised habitat that originally existed.

5. Develop researched approximations for how long it should take an imposed physical site alteration to translate into one or more specific ecological compromised conditions.

6. Through direct field observation and testing at the disturbed sites, and through comparisons with the reasonably matched reference locations and chemical stressor-only sites, identify and document instances of ecological compromise or upset at the former. Within a terrestrial context, a brief list of examples of compromise would be smaller reptile populations, smaller egg clutches, and erratic behaviors that cannot otherwise be explained. In an aquatic context, researchers might look for altered fish spawning and schooling behavior (perhaps due to munitions and other projectiles that interrupt stream and river flow), and an altered prey base for fish and birds (due to developing insect larva being challenged from a changed sediment particle-size distribution).

7. Related to guideline 5, develop researched approximations for how long it might take for the negative ecological influences attributable to physical habitat alterations to dampen out. This is important because it is rather certain that decades will have elapsed since the alterations were introduced. By way of example, consider if the absence of a certain tree species that fosters bird nesting and that formerly occupied a site is suppressing bird nesting activity at the present time.

8. It is imperative to understand that, well-intended as it might be, laboratory research can have no place in this study. By way of example, providing a physically disruptive environment in aquaria, and then monitoring swimming behavior or reproductive success with standardized test (fish) species, will not bring forth any utilizable information. (It is hoped that interested parties would well recognize the completely non-parallel nature of any artificial indoor arrangements and the real-world condition.)

9. It is imperative that interested parties also steer clear of experimental field studies. Again, these could not possibly parallel the real-world

condition (if for no other reason than the studies would be reporting on short-term responses at best). The study is not interested in bio-logical/ecological responses to newly created physical site impairments when physically altered sites (and the possible ecologically stressed site resources that stem from these) have existed for decades. For what it's worth, should the local ecology at artificially modified test sites be of a lesser quality than that of matched untampered-with control sites, such information will not inform for the topical study interest.

Alternative Approach Study Guidelines

1. Compile an initial list of terrestrial and aquatic sites with documented ecological impacts, i.e., where animals and plants have either notice-ably reduced vigor, or have depauperate populations. (Researchers are reminded that HQs greater than 1.0 do not signify, in the least, docu-mented ecological impact.)
2. Remove from the list all sites that have a chemical signature of the type that would warrant an ERA-type investigation (if not already undergo-ing one).
3. (Optional.) For each of the sites with the appropriate qualifications (i.e., pursuant to application of the two preceding guidelines), identify chem-ical stressor- and physical stressor-free habitat-matched locations.
4. Record detailed descriptions of the physical stressors present at each of the impacted sites.
5. Unless otherwise obvious, conduct the necessary literature research to reasonably assert that individual physical stressors are responsible for the observed reduced vigor and/or depauperate populations at the impacted sites, and provide the plausible, and hopefully proven, mechanism(s) for the impacts.
6. Provide an estimate for the percentage of sites with physical stress-ors where ecological impacts (as defined in guideline 1) are evident. Recognize that to do justice to arriving at the simple percentage sought here, an appreciable-size universe of appropriate-for-study sites should be reviewed. As with a good number of studies in this book, it follows that to bring about the best information for the topical area, multiple groups are encouraged to assume the study challenge.

Study Outcomes and Applications Thereof

The true intent of this study is to devise a means of assessing unaccept-able ecological risk—but, more realistically, demonstrated ecological impact—traceable to physically altered ecological settings for which man is responsible. Presumably, ecological risks and impacts due to physical stressors at sites (e.g., excessive scattered debris, inaccessible streams and other waterbodies that supply critical drinking water reserves, waste piles) can arise where these are the only stressors present, and presumably, too, this is why ERA guidance speaks of these stressors. It is most difficult to anticipate what the study outcomes could look like. Long before anticipating the percentage of trouble sites that have physical impediments accounting for lesser ecologies, or establishing linkages of physical stressors and their attack mechanisms, attention needs to be focused on the study sites. It is certainly not the intent to dissuade interested parties from stepping forward, but the merits of Alternative Approach study guideline 1 cannot be easily dismissed. It may be, then, that researchers will first find the task of assembling a list of ecologically impacted sites to be an insurmountable one. Should that be the case, the other exercises (e.g., assembling detailed descriptions of physical site alteration) become moot. As the reader surely knows, most contaminated sites are relatively small and, because of this, observing ecological impacts becomes a great challenge. Is a 10-acre property ecologically impacted because, over a discontinuous but cumulative half-acre of land surface, slag addition has pre-empted all plant growth and thereby photosynthesis? Does this 10-acre property, as described, support notably lesser bird or mammal populations than it should? (Perhaps ERA practitioners, regulators among them, are bothered by the sight of an altered landscape, but unsatisfied human wants [e.g., to not leave a compromised landscape as is] should not be dictating our actions.) With regard to the Primary Approach guidelines (i.e., where physically altered sites are identified, to then be investigated), readers surely again understand why ecological impacts might be entirely absent. While the strict science may dictate that physical impediments by themselves (let alone in conjunction with chemical stressor footprints) should lead to ecological impact, the time element cannot be ignored. There is every reason to expect that site receptors have responded to less than ideal site settings with the necessary adjustments to make due.

Study #29

Validate ERA Investigations for Military-Unique Contaminants/Sites

Premise

As contaminants to potentially be evaluated in ERAs, explosives, for all intents and purposes, are encountered by ecological receptors only in military settings. In substantiation of this claim, consider first that there are only so many fireworks manufacturers in the United States and that none of these come to mind as having severely stricken the local ecology through accidental spillages or other mishaps that may have occurred at their discrete locations. Consider, as well, that while there are numerous commercial manufacturers of explosives in the United States, (a) ecological habitat is unlikely to be rich and dynamic at the manufacturing plants and (b) this industry is tightly regulated with regard to the safe manufacture, distribution, storage, use, and possession of explosives. Other than spillages that undoubtedly occur at the plants (that do not warrant ecological assessment), opportunities for ecological receptor exposures in this milieu do not exist. In contrast to the above, it is the active military's purpose to develop and routinely test munitions, the latter releasing explosives (now included within the designations of "munitions constituents" [MC] and "munitions and explosives of concern" [MEC]) to the open environment. The releases, limited to discrete and remote locations (see Chapter 1), give rise to the question of whether a legitimate need exists for investigating MC and MEC in at least two capacities, namely the development of screening concentrations or benchmarks, and the estimation of ecological receptor hazard.

The intent of this study is to assemble the necessary information to support an informed decision on the legitimacy of including considerations for explosives in ERAs. A secondary intent is to articulate the consequences of including explosives in past, present, and future ERAs where the need for doing so is not demonstrated.

Study Guidelines

1. The areal scope of the potential problem area needs to be quantified. The following measures that could support a demonstration of insufficient need for ERA attention are sought:
   - the universe of military installations that historically tested munitions and/or presently conduct such testing;
   - for each relevant installation, the areal extent of land surface to sustain deposition of miniaturized explosive particles from the activities at a given range, factoring in, of course, wind-aided MC and MEC distribution (that can be either directly measured or reliably modeled);
   - for each relevant installation, the total acreage of usable/suitable terrestrial habitat, whether it be contaminated with MC and MEC or not;
   - for each relevant installation, the total acreage of usable/suitable terrestrial habitat that is unaffected by MC and MEC;
   - for a given state with one or more military installations that conduct munitions testing, a quantification of the percentage of total usable/suitable terrestrial habitat that is MC- and MEC-influenced.

2. Despite the challenges associated with sampling the soil of the downrange portions of munitions test ranges, representative MC and MEC concentrations for the biologically active zone (perhaps 0 to 15 cm bgs or 0 to 25 cm bgs) should be assembled. Note: published peer-reviewed articles reporting on explosives residues accumulations of the top one-quarter inch or top half-inch are not applicable. (Importantly for ERA purposes, the biologically active soil zone should never be influenced by the chemical(s) of interest. Animals and plants do not decide to utilize/venture into a certain soil horizons differentially in response to their recognition of the particular chemical suite/footprint that is present.)

3. Endeavor to characterize the small rodent community by trapping animals as close as feasible to downrange portions where soil MC and MEC concentrations are anticipated to be the greatest. Concurrently conduct small rodent trapping at areas of comparable habitat on the same installation that are (explosives) residue-free. For the two trapping sites, review trapping success (catch per unit effort), species richness, total captures, body weights, sex ratios, and overall appearance/health. Endeavor to apply the reasoning that if maximally exposed and areally limited small rodents are just as plentiful in the downrange areas as they are in the MC and MEC-free areas, training mission-related contamination is not contravening the health of other species. See guideline 5.

4. Endeavor to integrate into an installation's small rodent community analysis, observed instances of exceedance of published (health protection) soil benchmarks for explosives (e.g., for plants, invertebrates).

5. Only if the small rodent community was characterized, compute explosives-based HQs for downrange small rodents (this despite HQs not being risk measures). Relate the HQs to the outcome of the small rodent trapping/population analysis.

Study Outcomes and Applications Thereof

This uncomplicated study seeks to either validate or invalidate the purpose of forays into military-unique contaminant-based terrestrial ERA investigations at military installations. A simple review of affected and non-affected acreages might provide all the information to support a determination. The quest for a determination is seeded in the reality that opportunities to venture onto MC- and MEC-affected installation lands is severely restricted, and the prospects for regularly conducting assessments with site animals at the properties being highly unlikely. It would be prudent to seize an opportunity to demonstrate handily that all involvement with this category of site can be circumvented—relying, of course, on valid supports. Simplifying ERA, it should be recognized, is always to be desired, and one form of simplification is the dispensing with types of sites or scenarios altogether. Here, it is first recognized *a priori* that options to visualize animals in the field (for some form of health effects verification, for example) will almost always be precluded. Second, it is rather pointless to go through risk assessment motions at locations that are not slated to submit to cleanup efforts, but that are rather certain to see continued munitions testing whereby contaminant releases will contribute to still increasing concentrations of military-unique compounds. To the extent that the areal footprint of MC and MEC contamination on an installation basis or on a state-wide basis is minimal, such can facilitate a position paper on the lack of purpose of ERAs for military-unique chemicals. To the extent that the small rodent trapping reveals population health at downrange areas that is comparable to that of non-contaminated areas, such can facilitate a position paper on the non-need for ERA investigations for detonation range environments. To the extent that rodent work fails to unmask range impacts for this maximally exposed terrestrial group, the utility of developed laboratory-derived protective benchmarks for soil invertebrates and ongoing research in this vein will be thrown into serious question. A treatise following from such discoveries, one that paints soil invertebrate benchmarks for use at military ranges as fully unnecessary, could well allow for key inroads towards ERA reform.

Study #30

Validate the Small Rodent Hazard Quotients of ERAs

Premise

Hazard quotients (HQs) in terrestrial ERA intend to express the ratio of the assumed chemical-specific intakes of ecological receptors (for all intents and purposes, in units of mg foodstuff ingested per kg of organism body weight) to doses that are assumed to either be safe or effect-causing. For traditional food chains, HQs tend to be of greatest magnitude at the base, and this is quite understandable. The physically smaller forms of the chain base have substantially more intimate contact with contaminated soils, in part because they are areally restricted to the contaminated sites they inhabit by their notably reduced home ranges. Small rodents, veritable fixtures of terrestrial ERAs, demonstrate this phenomenon time and time again; their chemical-specific HQs, whether based on NOAELs or LOAELs, will commonly be an order of magnitude or more greater than the HQs of the species that prey on them, such as fox or owl. Theoretically, excessive rodent HQs at sites pose a grave concern, namely that the food chain's base is seriously uncertain, if not totally absent. Consider that no animal can be expected to survive if on a daily basis it is consuming one or more chemicals with toxic properties, at many multiples of either a safe dose or an effect-level dose. This review leads us to a great difficulty; rodent HQs in the tens and hundreds, signifying that these animals must be toxicologically overcome, are routinely crafted, but no contaminated terrestrial site for which such failing HQs have been computed has ever been documented as lacking its fullest complement of rodents. The situation clearly warrants investigation to explain how rodents can withstand what should be unavoidable reproductive impacts and outright lethality occurring via the ingestion pathway. Hence the subject study with its three distinct phases.

While ERA practitioners correctly peg the small rodent as forming the base or near-base of various terrestrial food chains, there are practical reasons for its selection as the focus of the proposed study. More than being relatively easy to work with for a host of reasons, the small rodent is real-istically the only routinely evaluated terrestrial receptor that can submit to the tenets of the study design. This reality well overpowers another, namely that, should it be thought to be imperiled, the small rodent does not register in man's psyche as an animal that can supply a site remedia-tion trigger. Importantly, should it turn out that small rodent HQs can-not be field-truthed (i.e., cannot be demonstrated to be actualized, which would in turn demonstrate that small rodents are *not* ingesting chemical doses that far exceed what should be safe), we will learn *de facto* that unac-ceptable HQs for all higher terrestrial food chain animals are also suspect.

Study Guidelines—Phase 1

1. As part of a desktop study, assemble a list of contaminated terrestrial sites, yet to be remediated, where HQ-based ERAs were calculated. Cull from that initial list sites for which at least one small rodent/insectivore was a receptor of concern. (There should be a limitless supply of these sites.) Cull from these a list of sites where the ERA terminated with either excessive NOAEL-based or LOAEL-based HQs for one or more rodents/insectivores. (There should again be a limitless supply of such sites.)

2. Construct a table such as the HQ Validation Matrix shown opposite and populate it minimally with at least ten rows of information. Endeavor to include more than one chemical from a site-specific ERA, and to populate the table with information from multiple ERAs. Note that the computa-tions leading to the populating of the table's last two columns is critical. Arrive at these column figures through necessary back-calculation, providing references for all values used in the calculations (e.g., soil-to-plant matter biotransfer factors). (Consider that a small rodent, such as a mouse, ingests about 8 g of food/day, and that approximately 2% of a small rodent's diet is soil.) Endeavor to apply the same assumptions used in the ERA from which the table-filling exercises are proceeding. Thus, the same soil-to-diet item biotransfers used in the ERAs should be applied for the exercise. Following from the table's first row of informa-tion, for the first of the table's estimated value columns, the "apparent" (average) soil concentration being sought is the lead concentration that should allow for a mouse having an estimated intake of 352 mg/kg/d.

3. Supply two additional columns for the table, similar to ones just described, but coinciding with that scenario where only that portion of the rodent diet that is soil (some 2%) is considered. Recall that in the case of herbivo-rous rodents (e.g., Meadow vole), with soil-to-plant matter biotransfer being generally quite poor, a soil ingestion-only scenario might generate very similar figures to those where a herbivorous diet is considered.

HQ Validation Matrix

Evaluated Receptor	Assessed Chemical	Toxicological Effect	Chemical Form Tested in Crafting the NOAEL or LOAEL	HQ Basis (NOAEL or LOAEL)	TRV Used (Body-Weight Adjusted)	Receptor's Estimated Intake (mg/kg/d; Taken From the ERA)	Excessive Site HQ (Taken from the ERA)	Estimated Values	
								"Apparent" Average Soil Conc. Corresponding to the Estimated Intake	Mgm Equivalents in the Volume of Soil a Small Rodent Ingests Daily
White-footed mouse	Lead	Reproduction (as reduced sperm count and motility)	Lead acetate	NOAEL	8 mg/kg/d (basis: 0.35 kg animal)	352	44	??	??

4. Extract from each ERA the (soil) exposure point concentration(s) used to generate the chemical-specific HQ(s). Compare these concentrations with the researcher-supplied (derived) concentrations for the table's last two columns. Highlight instances of gross incongruity between the concentrations.

5. Field-truth the exercise's estimated site soil concentrations (that should be giving rise to the ERA-reported excessive HQs).

6. Take several representative soil samples and measure (in mgms) the amount of chemical in the volume of soil that a small rodent incidentally ingests daily. Determine if chemical concentrations can account for the reported excessive HQs for the ERAs being investigated.

7. For several site chemicals, reliably estimate a site rodent's whole-body concentration. Collect approximately ten adult rodents of at least one (and preferably more) species and run whole-body analyses to ascertain the concentrations of those same chemicals. Compare the estimated and measured concentrations. (Note that soil metal concentrations should not have appreciably changed in the post-ERA years, and that there will be less certainty for organic compound-centered comparisons, recognizing that the latter can be expected to degrade with time.)

Study Guidelines—Phase 2

1. In the laboratory, adjust the diet (i.e., food, as opposed to water) of a common test rodent species with (preferably) several chemicals that gave rise to an ERA's excessive HQs, testing one chemical at a time. If at all possible, avoid administering the chemical by gavage, endeavoring instead to have the test animals (males and females) consume the chemical doses that an ERA (indirectly) indicates they have. Consider supplying only those food quantities that the test animal will reliably consume daily and that will deliver the intended chemical diet. (Note: chemical delivery by syringe can play no part in this study phase!) Have the test animals abide by the feeding regimen for a minimum of 90 days.

2. Periodically throughout the dosing period, observe animal behavior and monitor body weight.

3. Endeavor to assess dosed animals for evidence of the endpoints associated with the ERA's applied TRVs. Note that it is likely that reproduction, in one form or another, will have been the endpoint each time. Where (any aspect of) reproduction was an endpoint, proceed to mate dosed males with fertile females, and dosed females with proven-breeder males, monitoring reproduction with as many of the following as possible: percentage of successful matings, litter size, pup size and weight, pup survival.

4. Track the survival of the dosed animals along with their overall appearance and apparent health condition.
5. Where the TRV basis was survival (or longevity), compare the findings to the implications of the ERA's HQs. Where the basis of a TRV was one or more aspects of reproduction, compare the findings of the dosing study with the ERA's predictions.

Study Guidelines—Phase 3

1. Visit several sites of the type described in Phase 1's first guideline to become fluent with each's immediate habitat, to include even the more subtle of site features (e.g., topographical depressions of note, occurrences of rarer plant species, etc.). For each site, locate at least one highly matched non-contaminated area situated no more than a few miles away. (Note that, with regard to vegetation, sites should not only match with regard to the number of trees they support, but the types of trees should be factored in as well. Consider, for example, that nut-bearing trees draw the activity of certain rodents, and that a seemingly matching treed area that lacks nut trees can dampen the expectation that it supports the same rodent species and in the same numbers.)
2. Census the rodent populations of a number of sites and their matched non-contaminated reference locations, tracking, where possible, captures per unit effort. In a standardized way, track species lists for paired sites and record total small rodent captures, estimated population sizes, sex ratios, and age distributions.
3. Consider the collected population data in conjunction with professional judgment to determine if reproduction or any other toxicological endpoint (that served as the basis of one or more TRVs that contributed to notably failing HQs) has been achieved for any of the contaminated sites.

Study Outcomes and Applications Thereof

The study presents three separate opportunities for either validating or discrediting HQs, the measure upon which ERAs are so heavily reliant, although it does not estimate risk and amounts to no more than a meagre screening tool. With the literature recognizing that excessive HQs (particularly reproduction-based ones; to include NOAEL-based HQs of 20) are toxicological impossibilities, there is a need to know if the HQ can qualify as a reliable predictor. ERA practitioners would seemingly want to know this for an animal that is evaluated in virtually every terrestrial ERA, and one that can be evaluated up close. Importantly, the study's purpose is not to identify where computation-wise, HQs go so awry (given that rodents always seem to be plentiful at contaminated sites). The study's true purpose is to formally establish if HQs convey accurate or fraudulent information.

With regard to the study's Phase 1, the collected data will either show that rodents incur extreme chemical exposures or that they do not, with the whole-body analytics deciding this point. The collected and hopefully representative soil concentrations will allow for a check on the mathematics of the HQ computation, but cannot alone prove or demonstrate ecological impact. For the soil-as-diet review, an insufficiency of chemical to account for excessive HQs would put the finger on the gross imprecision of HQs, and allow for practitioners to hear that site rodents are not toxicologically challenged.

With the departures from the real-world condition acknowledged, the study's Phase 2 nevertheless allows us to learn if chemically exposed animals in the wild (based on the dosing of laboratory subjects with respectable parallels to what is said to occur at sites) are put at a deficit. The enthused researcher must recognize that despite the artificial/experimental design of the feeding study, the chemical delivery system nevertheless bests that of TRV-based studies many times over. Test animals that severely succumb with this study do not necessarily vindicate the HQ as a toxic effect predictor, although high survival rates and unimpaired reproduction outcomes will necessarily paint an unfavorable picture for the HQ.

The study's Phase 3, the most simplistic of the set, reminds us that, elementally, all that ever needs to be done in the world of ERA is to document that animals at contaminated sites are present as they should be (i.e., in proper numbers, etc.). Evaluating the small rodent community alone informs on the larger and unreachable species that often get evaluated although adequate evidence of their site presence is lacking. Sizeable rodent captures at sites that had excessive rodent HQs can only mean that HQs are spurious. Where rodent captures at sites parallel those of reference locations, an additional check on the system decides the matter. Enthused parties should understand that carrying out either or both of the study's first two phases is insufficient work; Phase 3 is absolutely needed.

Study #31

Defend the Inclusion of the Indiana Bat as a Receptor-of-Concern in Terrestrial ERAs

Premise

Many ERAs for terrestrial sites in the eastern United States include a consideration of the Indiana bat (*Myotis sodalis*). The heightened concern for this species is well understandable; it has been on the endangered species list since 1973, and has seen its numbers plummet by approximately 50% since that time. More recently, in the first four years since its discovery in 2007, the devastating white-nose syndrome has been responsible for the death of some six million Indiana bats. While interest over the well-being of the species is keen, certain aspects of its biology, in addition to concerns with several practical assessment approaches, would suggest that Indiana bat ERA treatments are fully unnecessary. The integrated review of these elements presented immediately below is advisable, and it will set the stage for the guidelines of this study.

- Indiana bats feed exclusively on flying insects, consuming their own body-weight of this living matter while foraging nightly over hundreds of acres. An ERA treatment for the bat would, at best, only entail a food model/HQ exercise, investigating contaminant transfer to the bat via its dietary uptake route. A consideration of the sizes of contaminated sites would contraindicate the need for the exercise. Unless a contaminated site—ideally a pesticide-contaminated one (for it is only this category of contaminant that has been implicated in the bat's reduced numbers)—extends to the hundreds of acres, a site's contribution to a bat's unhealthful dietary intake would be infinitesimally small.
- Indiana bats travel upwards of three miles from their day roosts to nighttime foraging areas. Often, though, in endeavoring to establish that Indiana bats are site-relevant for assessment, ERA ecological setting narratives dwell only on documenting the presence of suitable

day roosts at sites. Well-intended as these accounts are in establishing (perhaps) species presence, they unknowingly also furnish proof,
that (contaminant-bearing insect) dietary exposures are actually not
a threat; Indiana bats do not feed within the immediate vicinity of
their roosts. Thus, detailed site descriptions of dead and dying trees
with exfoliating bark, along with what the literature indicates are bat-
preferred tree diameters and densities, do not inform that contaminated sites themselves pose chemical threats to bats. It is more likely
the case that the descriptions are conveying information quite to the
contrary.

- Indiana bats cannot survive without appropriate hibernacula, and
these take the form of either caves or mines. Even then, for a cave to be
serviceable, it must have multiple entrances that allow for the regular
circulation of fresh cooled air. Just as the documentation of suitable day
roosts at sites actually comes to inform that chemical exposures are not
occurring onsite, so too the documentation of hibernacula at or near a
site is a strong support for a site not posing potentially harmful chemical exposures to bats. Upon exiting their hibernacula, Indiana bats
undergo a spring/late spring migration of considerable distance to the
regions where they will, for several months, exhibit their insect-foraging
behavior. Further, and with the best interests of the Indiana bat in
mind, those assembling ERAs would do well to recall that few caves
provide the conditions that facilitate Indiana bat hibernation. While
the bat's range extends to about 20 states, the ERA practitioner should
understand how patchy the distribution of functional hibernacula for
this species appears to be; by way of example, about 23% of this species'
total population hibernates in caves in the state of Indiana. Thus, it is
only a relatively few caves that are sustaining the Indiana bat population, which is estimated to be ca. 244,000 at the time of this writing.
Thought should probably be given to estimating how many Indiana bats
stand to be incapacitated from feeding on chemically contaminated
insects that stem from a typically sized contaminated site (perhaps 5
or 10 acres) located a considerable distance away from a hibernaculum.

This study invites interested parties to research the veracity of the
often-touted ERA concern that contaminated sites are impacting the
Indiana bat (which has already been imperiled from pesticide use and forest clearing). Pursuing a diverse array of tasks (i.e., the specified guidelines
below) should allow for the assembly of the information to determine the
rightful place, if any, of Indiana bat evaluations in ERAs.

Study Guidelines—Mapping Exercise (to Establish the Validity of Indiana Bat Inclusion in ERAs)

1. For each of the eastern U.S. states with Indiana bat populations, assemble the universe of hibernacula with appropriate air-flow characteristics and other requisite features to support hibernating bats. Plot the locations of the hibernacula on a map.

2. For each qualifying hibernaculum, construct the isopleth extending from 50 to 100 miles beyond it. (Note that the best existing data indicate that Indiana bat spring migrations are encapsulated by this band.) Plot the isopleths on the developing map.

3. For each relevant U.S. state (guideline 1), plot the locations of all federal- and state-listed pesticide-contaminated sites, managed under Superfund and Superfund-like programs, that (a) have yet to be assessed, and (b) have been assessed but have yet to have remedial work done. Depict the sites on the map.

4. Endeavor to establish that suitable day roosts (e.g., trees of some 29 species documented as being used by Indiana bats, that are of appropriate density at breast height, and have exfoliating bark) exist within approximately three miles of the identified pesticide-contaminated sites mentioned in the previous guideline. Where day roosts are sufficient in number (ideally plentiful), notate the three-mile band(s) on the developing map. Where day roosts are lacking or altogether absent in the approximate three-mile band encircling a site, remove the site from the map. Recall that Indiana bats forage, in part, at some three miles' distance from their day roosts. The absence of day roosts in the three-mile band effectively communicates that bats will not be feeding much if at all on insects that may have assimilated pesticides from contaminated site sources.

5. Endeavor to establish the migration distances of those insects that comprise the Indiana bat diet, and, where relevant, utilize this information in guideline 6.

6. Using the map layers that convey critical distances traveled (by bats— away from hibernacula and day roosts, and while foraging; by insects— away from the point-sources of sites from which they derive their pesticide tissue burdens), determine (a) how many pesticide sites could reasonably pose chemical threats to bats, and (b) the (contaminated insect) dietary fraction of Indiana bats whose foraging range might include one or more of the "qualifying" pesticide sites.

7. Produce a consolidated review on the subject of pesticide site exposure attribution to the ongoing Indiana bat population decline.

8. Using the assembled spatial ecology and contaminant exposure information, produce a report assessing the appropriateness and worthiness of including Indiana bat considerations, however simplistic they may be, in ERAs.

Study Guidelines—Food-Chain Exercise (to Establish the Validity of Indiana Bat Inclusion in ERAs)

1. For each valid site identified through the mapping exercise, estimate the number of insects and/or the collective biomass of insects bearing site-derived pesticides over an approximate five-month period (i.e., from spring migration in April through return to hibernaculum in September).

2. Estimate, with chemical uptake modeling (involving a consideration of lipid-based and other uptake/loading factors and transfer rates), the pesticide body burden of insects, but note that securing the body burdens by direct analytical measure (i.e., where composited insect samples are sufficiently large to satisfy analytical instrumentation requirements) is to be preferred. For the latter, ensure that insect collections are site representative (i.e., that they include a diversity of species [e.g., mosquitoes, midges, moths] that typify investigated sites).

3. Estimate the number of insects and/or the cumulative insect biomass that one Indiana bat might consume in a night, and also over an assumed five-month foraging period. Estimate the percentage of a bat's (insect) diet drawn from a contaminated site, given that nighttime foraging activity can extend over hundreds to thousands of acres. For this calculation, endeavor to apply a variable foraging area term for the five-month period, as the literature discusses same. As a conservative maneuver, have one calculation apply a fixed mid-summer (smallest) foraging area (of perhaps 10 acres) for the entire foraging period.

4. Calculate time-weighted site-specific reproductive HQs for the bat (i.e., those reflecting a reasonable number of foraging events [involving site-contaminated insects] occurring within an approximate five-month window), recognizing the gross level of uncertainty associated with this task. By way of example, while the TRV(s) used will likely be derived from one or more laboratory rodent studies, physiologically and toxicologically, a bat is not simply a flying mouse.

5. Endeavor to express the reliability of the HQs, and assess the value of the HQ exercise given the extreme level of uncertainty and the reality that in the common case, the Indiana bat will not be spatially relevant for (HQ or other) assessment at a typical contaminated Superfund-type site.

6. As a practical study end-product, researchers are asked to configure a guide that indicates the physical (e.g., areal) conditions that would need to be in place (e.g., size of contaminated site; relative locations of relevant day roosts) to trigger a legitimized contaminated site-based ERA for the Indiana bat.

Study Outcomes and Applications Thereof

While it appears that little can be done to remove the Indiana bat from its plight (owing to the continued use of pesticides and the continued clearing of forest cover in the United States), the study affords interested parties an opportunity to bring forward a level of analysis not yet assembled. It is likely that listed pesticide-contaminated sites do not individually (or even collectively) contribute in any meaningful way to the Indiana bat's reduced numbers. To the extent that this is true, we stand to learn, in a formal way, that pesticide sites pose extremely minimal threats to bats and/or that it is simply not feasible to scientifically weigh in on the potential of pesticide sites to pose risks to bats altogether. If these should be the outcomes, at a minimum it would follow that there is no point to (even) mentioning the Indiana bat (and other bat species experiencing similar population declines) in ERAs, to include "environmental setting" descriptions.

Engagement with this study broaches a larger area of neglect in ERA: that of the purpose and value of including any and all discussion on contaminated site elements where it is known *a priori* that definitive assessments for these are not possible. ERAs are not stronger when they include text that alludes to possible pathways of contaminant exposure that are not to be studied. Such text inclusions only ensure that the uncertainty sections of ERA's can be of greater length, and they annoyingly set the stage for not being able to fully close-out an ERA investigation. The writings to emerge from involvement with this study can potentially pave the way for having ERA practitioners appreciate that ERAs can often be markedly narrowed in scope. As an offshoot benefit of the newly gained awareness concerning the scope of ERAs, future ecological assessments stand to rightly focus their attention on core matters only.

Study #32

Defend the Inclusion of Birds as Receptors-of-Concern in ERAs

Premise

With regard to vertebrates, terrestrial ERA concerns itself with two phylogenetic classes only, Aves and Mammalia. (While some [for all intents and purposes, regulators only] might bother to argue the point, the reality is that there are no standardized methods for assessing chemical exposure risks to amphibians and reptiles. Further, it is unlikely that there are efforts underway to first introduce any evaluative methods for these two animal groups.) For each of these classes, ERAs conventionally list the species that are thought or known to be found within a site's immediate vicinity. Conventionally, too, one or more species are selected to serve as surrogates for other class members in what will, at best, take the form of a limited HQ exercise. Importantly, birds present their share of challenges within an ERA framework. First, there are limited available means to demonstrate that contaminated sites are imperiling them. Second, with contaminated sites generally spanning relatively limited acreage, anticipated bird counts for a given species are slight (and, dare we say from an ERA perspective, negligible?). Although exceptions abound, many bird species are migratory and thereby absent from contaminated sites for three or four months at a time, if not more. Such site absenteeism potentially provides birds with opportunities to clear contaminant stores they may have developed. Birds returning from their wintering locations may not reoccupy formerly used sites. In that case, chemical exposures may be limited to sporadic episodes occurring only over a portion of a year, episodes that are too limited to justifiably sanction ERA attention. Additionally, the smaller the size of a contaminated site, the less likely it is that a return-flight bird would come to occupy the contaminated site it did previously. Consider, for example, the prospects of a bird re-occupying the same four-acre property it did some five or six months prior to its southern migration of many hundreds of miles.

The subject study asks enthused ERA practitioners to draw on the existing voluminous body of literature on bird behavior and population dynamics, and in conjunction with a consideration of contaminated site features, substantiate bird assessments in ERAs. With the exercise intended to speak to the continental United States, multiple study groups are encouraged to become involved and to share their knowledge with one another en route to assembling the comprehensive defense being sought.

Study Guidelines

1. For each of the 48 conterminous U.S. states, compile by-county bird species lists, and then sort the county lists into functional (feeding) guilds.
2. Drawing on available documented field information and validated databases, assemble the following tabulated information for each species:
   - maximum density (as birds, pairs, and nests per acre or hectare). Notate where state-specific density information is available; average home range; average foraging range; average clutch size; average fledging success rate; the number of months/year when present in the county.
3. For each (by-county) guild, indicate the most likely species (no more than one or two) to serve as a surrogate in an ERA. Note: a genuine surrogate is one that is actually present at a site in question. Ensure that the tabulated information of guideline 2 is assembled for the selected surrogate species.
4. Based on available density and related information, indicate for each surrogate the maximum number of birds, pairs, and nests within 1, 5, 10, and 20-acre areas.
5. Provide a reasonable estimate of the total number of birds (i.e., across all species) that areally use 1, 5, 10, and 20-acre areas within each county at least 50% of the time that the bird is seasonally present (i.e., not when over-wintering at a distant location). (Recall two overarching and essential points. First, the overwhelming number of contaminated sites are only a handful of acres in size or less; hence the acreages specified in this guideline. Second, a key receptor-of-concern criterion is a species having a high degree of site fidelity; hence the suggested 50% area usage criterion. Consider that the smaller the site size [and the larger the home and/or foraging range], the less likely it is that a bird will satisfy the 50% site fidelity guideline requirement.)

6. For identified county-specific surrogate species (guideline 3), conduct bird and nest counts for multiple 1, 5, 10, and 20-acre pristine (non-contaminated) parcels. Coordinating with other interested research parties, arrange for the counts to be taken in multiple counties. Compare the field counts to the information gleaned from the tasks of guidelines 4 and 5.

7. Actively census multiple bird species (individuals and nests) at contaminated terrestrial sites that have not been remediated. (Presumably, these are sites where avian HQs greater than 1.0 have been computed and presumably, as well, where the HQ's toxicological endpoint was reproduction.) Compose a table with columns for (a) the unacceptable HQs, (b) the anticipated/predicted numbers (birds, nests) of the county-aligned surrogate species (corrected, of course, for site size), and (c) actual bird and nest counts for the surrogate species. Where possible, populate the latter two columns with count information (predicted and actual) for multiple species (i.e., species other than the surrogate).

Study Outcomes and Applications Thereof

This study stands to either validate or invalidate the practice of including birds in ERAs as a mainstay task. That birds are more easily and more regularly viewable in the environment than are amphibians and reptiles, and more amenable to dosing and tissue (e.g., blood) sampling than are these other groups, should not serve as bases for their regular inclusion. If birds should be as plentiful in contaminated sites as they are in non-contaminated settings, and especially where counts are collected at random times (the same times, though, for both site and matched reference location), there will be open demonstration that avian ERA concerns are not truly grounded. The study sets forth a two-pronged comparison scheme; site-specific counts versus literature bird densities, and site-specific counts versus those of nearby habitat-matched locations. The more instances of absent count differences in these comparisons, the easier it should be to advance improved thinking for ERA, much of it reflecting manifestations of contamination commonly being decades old. In particular, time-mediated opportunities for animals to adjust to chemically contaminated environs can be championed. There could also be great opportunity to suggest that censusing birds (to declaratively resolve the question of whether or not birds display appropriate/adequate site representation) should be the very first task to occur in an ERA or other ecological investigation. If a site is supporting the (bird) numbers it should—and particularly when the site condition clearly cannot become any worse—why should an ERA assess birds in a desktop or other fashion?! Importantly, a finding of adequate site representation promotes the notion that the presence of contamination (in site media or in living tissue) might often be a misleading site element; though we'd prefer the contamination not be there, it may well be impossible to show that it impedes any ecosystem component.

The censusing enterprise stands to heighten one other awareness, namely that because sites are often small, they govern an insufficiency of biota to legitimize ERA work altogether. As an example, if upon detailed field observation a site is found to be supporting only seven or eight birds (representing two species) a fair amount of the time, the worthiness of ERA effort investment is challenged. It is unrealistic to think that a site would be remediated for such a small number of birds, and particularly, of course, if that total number is in line with what the density-based literature has to offer.

Study #33

Validate the Contaminant Exposures of Granivorous Birds

Premise

Not uncommonly, ERAs assess (albeit crudely) more than one type of bird at a given contaminated site. Given that birds are variable in size, behavior, and niche, this could be a most responsible approach to assessment. When ERAs assess multiple birds with the standardized HQ approach, we know that they are primarily keying into different feeding designs. In a pronounced way, well-meaning ERAs illustrate their intent to be thorough in recognizing these variable designs when they necessarily include a (surrogate) carnivorous, granivorous, and omnivorous species. Curiously, while ERAs may convey the sense of there being an absolute necessity to include a bird of each distinct feeding guild, there is no strict requirement to do so. If there is any "rule" to follow when shaping the scope of ERAs receptor-wise, it is that ERAs should only include species that are site-representative. If, for example, a site does not actually support an omnivorous bird, such a bird should not be assessed. ERAs are certainly not supposed to dabble with hypotheticals, wherein stated conclusions could run along the lines of: "If this site had resident _____ (filling in the blank with an "omnivore"—for example, starlings), they would be protected because extremely low concentrations of site chemicals were detected in _____ (filling in the blank with one or a few diet items of an omnivore, perhaps here berries, insects, and earthworms)." There is also no excuse for including a bird in an ERA if it is hardly present or altogether absent from a site due to spatial dynamics; if a site, at best, draws no more than one mating pair (owing to the site's minimal size or the species' naturally low density), it's clear that the species should not be assessed. In the guise of being thorough through the certain inclusion of one bird of each distinct guild, ERAs often undo all the good they intend to provide. This occurs when the diet items of birds are not directly analyzed for their contaminant holdings. It is unfortunate that the species-specific nature of HQs to be computed stands to be lost through stopping short of collecting diet items that could be chemically analyzed. Where this occurs, appropriate birds for study at a site (chosen because they are, among other things, present in sufficient number to generate a plausible concern), will have their species-specific HQs only reflect unique body weights and assumed differential chemical intakes that follow from modeling exercises. In lieu of collecting species-specific diet items, modeling of the soil contaminants into the items proceeds. The potential for HQs to be misrepresentative are huge with this arrangement, particularly when we recall that birds commonly consume at least half of their weight daily. That variable site contaminants manifest differently in varied food items is another reality not being dealt with. A basic awareness, for example, that metals, as a group, tend to not be taken up by plant roots, or to be translocated from the roots to above-ground plant parts, is a case in point. Where site/soil contaminants are predominantly or exclusively metals, direct contaminants measurement in the bird diet is a rather indispensable task for an assessment if it is to have any true value.

This study's focus is granivorous (i.e., seed-eating) birds, a distinct category of bird considered often enough in terrestrial ERAs. The essential study component is straightforward: to collect seeds at contaminated sites so that their contaminant residues can be measured. A consideration of spatial realities, as we should expect, is another layer of analysis woven into the research scheme.

Study Guidelines

1. As should be obvious, this study avails itself to multiple interested parties. The meaningfulness of findings is always augmented with larger datasets, and the possibility of identifying defensible and utilitarian trends arises out of these. The reality is that limited budgets will preclude individuals from traversing the United States to acquire a broad array of samples of variable environments. Thus, a consolidation of findings reflecting a greater geographical range may only emerge where researchers, *a priori*, intend to collaborate.

2. Assemble a list of contaminated terrestrial sites that are either slated to go through a risk-based remedial process, or that are already involved with one, where seed-eating birds have a regular noticeable presence, and where the total bird count is at least ten. Only populate the list with sites that have not had soils remediated. Only populate the list with sites that distinctly support vegetation that produces seeds that onsite

granivorous birds regularly consume. Note: a site would not be viable for the study list if granivorous birds, though demonstrating high site fidelity, forage beyond a site's boundary to acquire dietary seeds (as might occur should the focus vegetation not occur onsite).

3. Collect soil samples of variable and discrete depths at the very locations where seeds abound. Specifically collect surface soil, and soil at root depth. (Note that the latter requirement necessitates knowledge of the root depth of the seed-producing plants.) Have the soils analyzed for those contaminants that contributed to the site's formal placement onto some agency's registry of hazardous waste sites (such as the NPL).

4. If anticipated present-site contaminants are detected and occur in concentrations professionally judged to be ecologically concerning, collect sufficient seeds to constitute several samples, and ship these off for laboratory analysis. Analyte lists should be comprised of the site chemicals that trace to historical site use.

5. Tabularize sample-specific seed analyte concentrations with collocated soil analyte concentrations (for both surface and root-depth soil). Compute bioconcentration factors for all seed-detected analytes.

6. Using the laboratory-detected seed concentrations (guideline 4), compute HQs for one or more qualifying site granivorous birds (see guideline 2). If an ERA of some kind was already done for a bird (preferable a granivore) at the site, endeavor to use the same chemical-specific NOAEL and/ or LOAEL that was previously used. Where site chemicals are detected in sampled surface soil, include a 10% dietary component of surface soil in the computations. If site granivores were assessed differently, mimic the previous work, adjusting only the contaminated seed term.

7. Where previous HQs and those of the study are similarly generated (with this primarily referring to the exposure assumptions used), provide a side-by-side HQ comparison. Endeavor to account for any observed HQ magnitude differentials.

8. Consider returning to sites to collect seeds at a second time of the year (assuming such are available), submitting these again for contaminant analysis, to be followed by HQ computation. Return to guideline 7.

Study Outcomes and Applications Thereof

Where chemicals have been sequestered in soil for 30 or more years, which is always the case in terrestrial ERA, chances are good that the local ecology has adjusted to the chemical presence, if it even needed to. Should seed-eating birds have a respectable presence at a given contaminated site that could qualify for the subject study, in truth the study is not needed— at least as far as helping us gain an understanding of the potential for a contaminated seed diet to pose harm. Why else are we seeing the (granivorous) birds at the site presently?! There remains, nevertheless, great merit in researching the veracity of the granivore dietary concern, doing things the right way. It might be asking a lot of a plant to shunt soil contaminants into its seeds, and to do so continuously in a fixed location (for plants do not move) for multiple decades. While we can perhaps appreciate that side-stepping the direct measure of contaminant holdings in seeds makes for a time- and a money-saving measure, modeled seed contaminant levels come at a cost, namely a sizeable uncertainty associated with just how well they align with actual seed levels. Directly measuring contaminant levels in seeds and using these determinations to fuel dietary HQs can shed great light on these computations. What if HQs based on modeled seed concentrations consistently fail, but HQs based on actual seed content measurement "pass" or are of vastly lesser magnitude than the modeled-based HQs? The consolidated information to be acquired with this study, particularly where this reflects a multitude and diversity of sites (from the standpoints of geographical location, type, and number of contaminants, contaminant concentration ranges, site age, etc.), could lead to the discovery that the distinct category, that is, the granivorous bird, need no longer be ERA-assessed. Of course, such a discovery can only be appreciated where ERA practitioners understand that ERA simplification (as in demonstrating that a given assessment component need not be evaluated any longer) is a most desirable gain. Perhaps, study findings will reveal that for (only) certain contaminant classes (e.g., metals), the seed ingestion pathway falls way. This alone would be a saving; in advance of otherwise including granivorous birds in ERAs without thinking, we would then be equipped to recognize when certain assumed receptor-of-concern species are inappropriate for assessment inclusion.

The granivorous bird was singled out for this study because of its distinct diet and with the recognition, as was mentioned before, that a substantial portion of the diet is comprised of seeds. While we cannot know of the degree to which a site's contaminants load to seeds until data is gathered, there are good reasons to suspect that seed uptake is markedly low. Enthused researchers will, even with this reading, anticipate what is next suggested as an ERA research task (although not formally described as one in this compendium). A concerted effort to establish the actual concentrations of site contaminants in the diet items of other common ecological receptors (e.g., mammals), and not just birds with still different feeding designs, would be highly utilitarian. At the very least, we would be equipped with vastly improved dietary item concentrations to support the regular modeling that occurs (although such modeling unfortunately contributes to the anemic and non-desirable HQ construct). At best, we may come to discover that other receptors for certain sites and contaminant suites also need not be included in ERAs. In other words, the described work stands to uncover other opportunities to simplify ERAs—a phenomenon most welcome.

Study #34

Survey ERA Practitioners with the Intention of Uncovering Faulty Understanding and Bias

Premise

Worthy studies designed to bring about improved ERA can take on many forms. To the extent that this compendium's offerings are rightly grounded, the multitude of topics presented attest to the numerous areas where misinformation or altogether lacking information abound. Critically, the reader should not think that the locations where the strides need to be made for advancing ERA science are limited to the laboratory bench and the field setting. Thus, an over-emphasis placed on securing (what we think to be) more accurate TRVs, or a bettered understanding of the percentage of an animal's diet that is incidentally ingested soil, will not get us there. Studies that seek to get at the types of information we routinely work with, but where it is of a more refined nature, could be external trappings that are best avoided. Perchance, our basic thinking on ERA is questionable and because we don't bother to investigate certain possibilities (e.g., that, truly speaking, site animals couldn't possibly be at risk from chemical exposure), we instead proceed with trying to make the design that the ERA field has for all intents and purposes settled in on work better.

This study, potentially the most ambitious of those described in the book and seemingly the most different from all others, is suggesting that, for ERA to come into its own, elements of the field that we have nonetheless taken to be givens (e.g., that there is still a need to assess ecological risk at a multiple-decades-old contaminated site) need to be cautiously scrutinized. Long before we consider such things as how many samples to collect, and how to interpret the responses of test animals that were administered chemical doses, we need to become more familiar with why we have engaged in ERA work at all. At a minimum for the Superfund-type site arrangement (this book's concern) we need to take a major step

back, or, more realistically, quite a few of them, to understand the thinking behind what we do. This study invites interested parties to probe at the
psyche of the ERA practitioner. There are two available avenues for acquir-
ing the highly utilitarian information being sought—conducting scientific
surveys, and questioning subjects placed under hypnotic trance—and it is
recommended that both be pursued. Interested parties who are drawn to
one avenue over the other might elect to team up with professional peers
who are of another bent, and for whom the alternate avenue presents a
more curious challenge and a grand(er) opportunity. In either case, the
same types of individuals to serve as test subjects are being sought. These
are people who in some way facilitate the ERA process. These should
include those who pen ERA guidance, those who assemble ERAs, those
who review and comment on ERAs, and those who negotiate with other
stakeholders about remedial action possibilities pursuant to completed
ERA work.

Study Guidelines: Scientific Surveys

1. It goes without saying that only qualified professionals with specialized expertise in the design and administration of scientific surveys
   should be sought out. To the extent that the intent is to have ERA practitioners reveal misunderstandings in the essential science of ERA
   and any biases they might harbor, particular care must be given to the
   sequencing of questions asked. Indeed, the survey should necessarily
   be designed to entrap test subjects if at all possible. By way of example,
   the studious ERA practitioner understands that (a) HQs do not express
   risk, (b) so-termed "unacceptable" HQs are not proof enough of ecological damage or the potential for it, (c) unacceptable HQs do not provide sufficient information to trigger taking remedial actions, and (d) it
   is unthinkable to move for taking remedial action without first visiting
   a site. Where the survey respondent has indicated that he/she negoti-
   ates cleanup decisions as part of his/her work, it is imperative to know
   if such an individual visits the very sites that constitute the negotiable
   subjects. To ask survey respondents if they visit sites *after* asking them
   if they negotiate the fate of sites (and where they have answered in the
   affirmative), would likely prompt respondents to lie on the latter question. Respondents would not want to convey that they need be armed
   with nothing more than desktop (i.e., HQ) information to declare sites
   ecology-compromising, and respondents would realize that avowing
   that they do not visit sites would be incriminating for them. As a safe-
   guard against respondents changing their responses so as not to reveal

that they employ poor practice (as they may now first realize), surveys should be designed to not allow respondents to revisit questions already answered. The extent to which this deliberate-design survey limitation may impact on the value of the data gathered, if indeed it impacts at all, can be dealt with at a later time.

2. In an overall sense, the survey should be designed to have respondents (unknowingly) indicate that their (strict) adherence to the applied ERA process overshadows their thinking about the actual site ecology.

3. As to amassing a population to survey, it would make the most sense to tap professional environmental/ecological societies and agencies, or subsets of these. Surveys should be administered electronically, with an alias name used as the point of contact. As evident in guideline 6, those taking the survey are asked to identify their place in the ERA field. It is suggested that survey questions and the sequence of these be adjusted to conform to the variable roles of the respondents (e.g., contractors versus regulators). The responses to the first and second questions, then, should trigger the delivery of a series of survey questions that are tailored to a specific category of respondent (e.g., ERA reviewer of a non-regulatory stakeholder agency).

4. Being completely up-front about the intent of the survey (i.e., explaining that the objective is to learn of the numbers of ERA practitioners who don't well understand the ERA process and who fail to see its considerable shortcomings) would rather obviously contribute to a fully worthless effort. Unavoidably, it is therefore imperative that the survey's introduction be a fabrication, perhaps explaining that the point of contact is a graduate student who is interested in putting forward a white paper on developing new ERA guidance that reflects the knowledge, skill sets, and practices of ERA professionals. The reader should duly note that psychosocial science is often reliant on some order of fabrication in order to procure the data being sought. (Thus a study about anxiety might necessitate test subjects being told that there is an attack dog in the room to their right, and a kitten in the room to their left, although none of this is true.)

5. Surveys should necessarily include fill-in questions. These allow for obtaining valuable information without having to resort to a testing mode, something to be avoided given that, traditionally, survey-takers do not take well to being put on the spot. To illustrate, the survey will want to discover how prevalent it is that ERA practitioners believe the HQ to be a risk measure. Rather than list out several terms as choices (with the HQ as one of these) and ask "Which of the following are risk measures?," it would be much better to ask: "What risk measures do you find to be most helpful?," leaving the respondent free to provide his/her own offerings.

6. A roughed-out survey would ask the following:
   - Identify yourself professionally through your ERA involvement:
     regulator _____ government employee _____ contractor _____
     industry/private sector _____ academia _____
     natural resource trustee (e.g., Dept. of Interior) _____
     natural resource trustee supporting agency (e.g., U.S. Fish and
     Wildlife Service) _____
   - Indicate the extent of your ERA involvement vis-à-vis contaminated
     sites:
     ERA author _____
     reviews and comments on ERAs (i.e., ecological risk assessor) _____
     negotiates with other stakeholders about remedial activities pursuant
     to completed ERAs _____

Specific Survey Questions: (Select the Best Answer to Each)

1. With regard to your negotiating site remedial strategies, what percentage of sites do you visit?
   a) 0%   b) up to 25%   c) 25 to 50%   d) more than 50%
2. On average, how long do your site visits last?
   a) 1–3 hours   b) 3–5 hours   c) 2 or more days
3. On average, how many visits do you make to a site where you are involved with remedial negotiations?
   a) 1   b) 2–4   c) 5 or more
4. How often do you observe field evidence of the harmful effects of chemicals on ecological receptors?
   a) never   b) rarely   c) moderately   d) frequently
5. Would you recommend/require site cleanup on the basis of contaminated environmental media (e,g., soil) only?
   a) yes   b) no
6. Would you recommend/require site cleanup on the basis of one or more unacceptable HQs?
   a) yes   b) no
7. What magnitude of hazard quotient (HQ) confirms for you that a given site ecological receptor is at risk?
   a) >1   b) >5   c) >10   d) >20
8. Regarding the sites that you work with (in one capacity or another), generally how long have they been contaminated?
   a) 5–10 years b) 10–20 years   c) 20–30 years   d) >30 years
9. How many individual foxes should be present at a site for it to be rightly designated a receptor-of-concern?
   a) 1   b) 2   c) 5   d) 10 or more

10. What minimum site size legitimizes the selection of the fox as a receptor-of-concern worthy of assessment?
    a) 1 acre   b) 2–5 acres   c) 5–10 acres   d) 15–20 acres
11. What minimum site size legitimizes the selection of the white-tailed deer as a receptor-of-concern worthy of assessment?
    a) 1 acre   b) 2–5 acres   c) 5–10 acres   d) >10 acres
12. What minimum site size legitimizes the selection of the robin as a receptor-of-concern worthy of assessment?
    a) 1 acre   b) 2–5 acres   c) 5–10 acres   d) >10 acres
13. What minimum site size legitimizes the selection of the red-tailed hawk as a receptor-of-concern worthy of assessment?
    a) 1 acre   b) 2–5 acres   c) 5–10 acres   d) >10 acres
14. Do HQs above unity (1.0) for site earthworms indicate that vermivorous (worm-eating) site birds are at risk?
    a) yes   b) no
15. Rank—from strongest to weakest—the following lines of evidence for supporting the case that there is unacceptable risk at a terrestrial site.
    1. a failing plant toxicity test
    2. HQ of 5 for a small rodent
    3. presence of xenobiotics (e.g., PCBs, TCE) in the soil
    4. site contaminants present in liver (in a bird or a mammal)
    5. areas of stained soil visible
16. Rank—from strongest to weakest—the following lines of evidence for supporting the case that there is unacceptable risk at an aquatic site.
    1. a failing elutriate toxicity test with an invertebrate test species
    2. a failing elutriate toxicity test with a fish test species
    3. exceedance of one or more sediment quality criteria
    4. an SEM/AVS ratio greater than 1.0
17. What is the literal meaning of a NOAEL-based HQ of 7? What is the literal meaning of a LOAEL-based HQ of 4?
18. Is there a difference between an excessive HQ at a site that recently became contaminated and one that has been contaminated for 30 years?
    a) yes   b) no
  If yes, briefly describe the difference.
19. Is a site-chemical body burden a condition to correct?
    a) yes   b) no
20. If it should be that only one animal of a given species is potentially harmed at a site, is this a problem?
    a) yes   b) no
21. Can one or more "failing" HQs at a contaminated site provide the basis for taking remedial action?
    a) yes   b) no

22. How should site ecological receptors (e.g., meadow vole, red-tailed hawk) with NOAEL-based HQs of 20 for the reproductive endpoint physically appear?

23. Have you ever participated in site-specific ERA (cleanup-centered) negotiations?

     a) yes   b) no

  If yes, have you ever moved for cleanup when there was evidence of visible impact (harmed animals, depleted populations that were traceable to a chemical release)?

24. Do HQs above unity (1.0) for a site's small rodents indicate that site rodent predators are at risk?

Study Guidelines: Hypnosis

1. Recognize that securing test subjects will undoubtedly be the most difficult task of this study portion. Test subjects are of two kinds; those who write and/or review ERAs, and those who set ERA guidance and/or negotiate remedial actions. Regarding the latter, an effort should be made to secure, as test subjects, individuals employed by the U.S. EPA.

2. In soliciting for participants, advertising should explain that hypnosis will be a study method and that only licensed professional hypnotists will be assisting. Only individuals who provide a signed consent form can be participants. Advertising should also mention that study participants will be paid for their time.

3. Although some of the questions to be administered in this phase of study are of the "yes/no" type, many questions should necessarily seek out narrative responses. With such responses, test subjects are not confined to selecting "fixed" choices, and the potential for ambiguity in responses is markedly reduced. All question-and-answer hypnotic testing should be recorded.

4. Below are some core questions to use. Interested researchers are free to assemble their own battery of questions, and to sequence them as they see fit. Researchers should be answerable for the actual arrangement of the questions they use.

Core Questions

0. Is there potential for (ecological) risk at a site that has been contaminated for three or four decades or more?

1. Is the hazard quotient (HQ) a measure of risk?

     a) yes   b) no

2. What is the ecotoxicological meaning of a HQ of 10 for a reproductive endpoint?

3. What is the ecotoxicological meaning of a HQ of 25 when mortality is the endpoint?

4. Can ecological receptors with HQs of 100 or more still populate a site?

   a) yes   b) no

   If yes, briefly explain.

5. Is there a minimum size requirement for a site to submit to ERA in some form?

6. After about how many years of a site bearing contamination should one expect to see ecological impacts?

7. After how many years of a site bearing contamination is it too late to be assessing risk? Explain your answer.

8. About how many red or gray foxes occupy a 2-acre land parcel?

9. Is there purpose to conducting plant toxicity testing for a site that is well-vegetated?

   a) yes   b) no

   If yes, why so?

10. Is a failed earthworm toxicity test (based on a contaminated site's soils) good enough to conclude that site remediation is necessary?

    a) yes   b) no

11. If, at a site, only small rodent HQs were deemed unacceptable (i.e., and not for any other terrestrial receptor), would it be appropriate to consider remediating?

12. What is the ecotoxicological meaning of a HQ of LOAEL-based HQ of 30?

13. Is a "final HQ" a screen or a definitive value?

    a) yes   b) no

14. Are HQs above 1 sufficiently reliable to indicate that site ecological receptors are having harmful chemical exposures?

    a) yes   b) no

15. If censusing and other forms of field observation show a contaminated site to support the species that it should, and in appropriate numbers, might the site still need to be cleaned up?

    a) yes   b) no

    If yes, why so?

16. In terms of identified unacceptable risk, does it matter how long a site has been contaminated?

    a) yes   b) no

17. Is a HQ of 20 twice as threatening as a HQ of 10?

    a) yes   b) no

18. What magnitude NOAEL-based HQ is an impossibility? What magnitude LOAEL-based HQ is an impossibility?

19. Name two natural resource trustees.

20. What organisms are likely at risk if the sediment of a contaminated freshwater system produces failing toxicity test results with an amphipod crustacean as the test organism?
21. Does sediment remediation work proceed for the purpose of protecting benthic macroinvertebrates?
    a) yes   b) no
22. Does the presence of contamination in a medium necessarily mean that remedial action needs to be taken?
23. Could you vote or negotiate to leave a site "as is" although it still bears contamination in its media?

Study Outcomes and Applications Thereof

Conducted appropriately, this study has enormous potential to reveal the errant thinking of ERA professionals as regards the ERA process applied in contaminated site work, something vitally needed. There is no shortage of misunderstandings in the science that prevails, and there are truisms that are forever overlooked. Thus,

- there are no measures of ecological "risk" (i.e., expressions of the probability of a receptor developing a toxicological endpoint);
- NOAEL-based HQs as low as 20 are impossibilities;
- it is pointless to put miniscule (aquatic or terrestrial) sites (e.g., those covering an acre or two) through even the most rudimentary of ecological evaluations, or to conduct any work at multiple-decades-old sites.

An anticipated outcome is that survey respondents will tenaciously cling to and defend ERA process, while those under hypnosis will tell a very different story. Of course, with the cautious sequencing of survey questions, survey respondents can be expected to trip themselves up often enough, as in admitting that remedial decisions can be made without once venturing onto site property. It is anticipated that hypnotized subjects will acknowledge any or all of the following:

- that small sites do not realistically submit to ERA investigation because they do not support enough biota;
- that it is pointless to clean up sites to protect earthworms or small rodents;
- that toxicity tests are highly misleading, and that the ERA process does little more than serve as a place-holder, supplying a seeming parallel to the human health risk assessment (HHRA) scheme for non-human species.

To the extent that the anticipations are correct, the necessary data will have been collected to demonstrate that ERA investigation often unnecessarily proceeds and that a process for the conventional contaminated site arrangement is altogether unnecessary. Provided that bias does not interrupt the dissemination of the data to be forthcoming through peer-reviewed networks, astute influential and forward-thinking political leaders can engage in implementing environmental assessment reform that is so badly needed.

Denouement

A most welcome coincidence… Properly, I have postponed the writing of this chapter that is to supply additional context for the foregoing 30-odd described studies until all of those have been written up, cogitated upon, and had final tweaks supplied. At the very time that I sat down to assemble my thoughts for this concluding chapter, I encountered not one, but two situations that amazingly validated my purpose in writing. They are described briefly below. They make clear that there is a pressing need to take ERA out of its aimless state and to give it direction, and the studies that have been presented were intended to assist in that area. The two situations remind us that some ERA practitioners can display an overextended, if not blind, allegiance to the process in place, while others can elect, perhaps without basis, to not apply any of the process. Perhaps I am being too kind, and it is not so much a conscious allegiance to the current process that perpetuates poor science happening in ERA investigations, but rather a certain pervasive mindlessness. There are surely practitioners out there perfunctorily going through the motions, thinking that they are doing good work because they are doing just as the guidance instructs. What, though, if the guidance is in error?

The first situation had a brief deliverable concerning ecological calculations forwarded my way. Before opening the file, I wondered what calculations these would be, for there are no calculations in ERA—other than the hopeless ratios we know to be HQs. In fact, the calculations *were* HQ-related; for a terrestrial site I knew nothing about, there was a series of neatly formatted tables, each illustrating how (supposed) "protective concentration levels (PCLs)" had been computed for a fixed series of four receptors (two birds and two mammals), with a separate page allocated to each of seven metals. In the event it wasn't clear what the work was showing, atop a box containing the HQ equation at the start of each page was a line of type: "PCLs are soil concentrations back-calculated for HQ = 1.0." PCLs, then, were simply preliminary remediation goals (PRGs); the contractor had elected to own his/her work through introducing an acronym slightly different than the one usually used. A two-page read-ahead accompanied the tables. It explained that site activities had resulted in soil-metal accumulations of concern, and it supplied the site size, some 17 acres. For what should be obvious reasons, I was certainly not going to review the calculations. My thoughts ran to composing a brief memo that would politely explain that there would be no purpose served in checking over any (back-calculated HQ) math associated with tabular information on food ingestion rates, dietary composition, and more. With diplomacy, the memo I envisioned would articulate that, in contradistinction to the tables that had been perfunctorily generated, there was a set reason—a conscious and mindful one—for my not bothering to look over the supplied numbers. Thus, the memo would explain that the assembled tables were exceedingly premature, i.e., that a site does not simply advance to the stage where cleanup levels (or "protective concentration levels," as the case may be) are crafted until there has been demonstration of a site condition in need of repair. "Protective concentration levels" connote that the site, as is, exposes ecological receptors to chemical concentrations that are obviously not protective. But was that true? I should mention that, for the mammalian component of the PCL effort, the two species considered were the white-footed mouse and the short-tailed shrew. Were there no larger mammals utilizing the 17 acres of land? Had I ever come across a situation where the developed mammalian PRGs pertained only to rodents? I halted articulating the familiar cavalcade of troubling issues I would ordinarily bring up in a situation resembling this one. Just a smattering of these include:

- Does anyone clean up a site so as to afford protection to (just) small rodents?
- Are there no small rodents occupying the 17 acres of pine-hardwood forest that constitutes this very site?
- Have there been any accounts of a decimated rodent population at this location?
- Regarding the back-calculation, is the deliverable's author not aware that HQs are not linearly scaled?
- And the "classic" issue that every engaged ERA practitioner should know to champion: "Now, after how many decades of soil-metal contamination, there is a push to remediate?"

My e-mail response, in lieu of a review memo, was brief and to the point. It said that the deliverable was not reviewed because the approach taken was fully inappropriate, it reflecting a non-familiarity with ERA's intent. The e-mail triggered a rapid, dramatic, and promising result. In just hours, I received a call from the program manager, who informed me that he and his group felt the same way (i.e., that there was no reason to even suspect that ecological species at the site were being impacted), but that they were inhibited to articulate this position. (The reader is asked to recall the book's preface, wherein the hope is expressed that ERA practitioners, emboldened through their acquisition of new and superior technical information, shed their timidity and develop the confidence to call out what they observe to be just plain wrong.) I have yet to mention where the PCLs deliverable was headed, but this is as good a place as any to do that. In order to achieve the developed PCLs, the intention was to have the top two feet of soil across the 17 acres removed, to the tune of $12 million.

My subsequent consultative input to the project rang true in the ears of the parties involved. The preparers of the misguided deliverable had overlooked the reality that, simply because contamination is present, concluding that ecological receptors are at risk and that a cleanup must proceed do not automatically follow. Further, there could be appropriate testing applied to establish if the receptors, painted as imperiled, were really so. Critically, a negotiating party can reserve the right to demonstrate (or, perhaps more appropriately, have a site demonstrate for itself) that, despite contamination, all the species that should be supported, in fact are. At the time of this writing, the site is to submit to direct health status assessment for rodents (see Study #9), a maximally exposed animal group that the deliverable would have the parties believing is incapable of reproduction (many times over!). Soon enough, we will know if the site's small rodents are reproductively impaired or perfectly fine. And the take-home here is that with situations like this one, not at all uncommon, so great is the need to develop additional direct health status assessment methods akin to RSA.

The previously described situation had parties who were only too happy to haul dirt on behalf of ecological receptors, but notably without a shred of biological data from the field to support their intentions. The second situation presented as a polar opposite to the first, and it continues to leave me stymied. A terrestrial site of 13 acres that previously supported the industrial processes of metal working and finishing operations bears degreasers, oils, solvents, electroplating solutions, and sludge as the primary wastes emanating from those past activities. The contaminants-of-concern list runs to a series of metals in high concentration, volatile organic compounds, semi-volatile organic compounds, and polychlorinated biphenyls, among others. As to the site description, it is a mostly undeveloped, regularly mowed and treeless grass-covered field. For this site, the powers that be were writing off ecological concerns entirely, stating that the site supports no terrestrial receptors. The oddity here, in my perception, is the responsible party's resolve to put forth the notion that the site requires absolutely no ecological consideration. ERA practitioners are undoubtedly familiar with regulators rejecting any such motions to bypass ecological review, regardless of circumstance. If regulators lash out about sites that are truly miniscule (perhaps covering no more than one-tenth of an acre), it's clear what they would have to say about our profiled 13-acre site. Although the science supporting the dispensing with even simplistic ecological reviews is rooted (e.g., there simply are no populations of interest present at sites that occupy two-tenths of an acre), a regulator won't want to hear of this. How, then, did this non-asphalted site manage to (thus far) pull off such an apparent victory? I say "victory" because, to my mind, every opportunity to circumvent the computation of HQs should always be seized, because HQs are not at all helpful. Readers whose job tasks include regularly reviewing ERAs might want to ask themselves if they ever recall a site as large as this one not having had to submit to some degree of ERA treatment.

Perhaps the reader can anticipate why I say I am stymied in this instance. How confident are we that this site supports no ecological receptors? Of worthwhile note, information supporting the allegation of absent ecological receptors was not furnished in the report I received. Were there no birds alighting on the ground over the entirety of the 13 acres—birds that could feed on contaminated worms and grubs, and incidentally ingest bad soil in so doing? Were there no mammals occupying the 13-acre space at least some of the time? To the extent that our site presents as something akin to an uneventful golf course, is it a fact that golf courses are never subject to ecological assessments? Were this to be true, we would have our foot in the door for trying to increase the minimum contaminated site size necessitating some form of ecological review, something that would seemingly be of great interest to many practitioners. Presently, we know that a handful of states have weighed in on this matter, and it appears that (with some caveats built-in) three acres is about as large as a contaminated land parcel can be without having to submit to ERA.

The essential take-away from the two situation descriptions should be clear. ERA is in a proverbial free-for-all; there is no consistency in how ERA is applied, undoubtedly reflecting that the discipline's very purpose is not truly understood. As you have now read, some will proceed to cleanup considerations with no data to support a site presenting with manifested ecological impacts or valid indications of these forthcoming. Others will elect to completely side-step ERA review for a parcel as large as 13 acres. Those fully immersed in ERA work will surely find the latter case to be particularly disturbing, given the commonplace soup-to-nuts ERA treatments rendered to sites that occupy all of one-tenth of an acre or even less.

Oddly enough, in order to capitalize on the information the foregoing studies stand to bring forward, a treatment on the intent of ERA as applied at conventional hazardous waste (Superfund-type) sites is warranted—even at this very late point of the book. If the human health risk assessment (HHRA) determines humans exposed to contaminated media for their likelihood of developing serious toxicological endpoints, then ERA should strive to do the same for non-human species. It is as easy as that. While there is certainly more to ecology than merely sizing up the health state of certain species inhabiting a location, ERA is not cut out for higher-order investigation and analysis, and it never was. Importantly to this day, ERA has never ventured into anything more sophisticated than organism assessment, and that being the case, why should it try to take on something more sophisticated now? Consider, too, that ERA's track record where it concerns only the organism/effect aspect receives no raves; HQs do not inform in the least about the health of ecological receptors. In the past, cleanups may have proceeded based on HQs, but no one can point to demonstrated ecological impacts in the field that coincided with elevated HQs.

Following from the above, ERA's essential tasking becomes one of assembling a site's list of organisms that can be said to legitimately submit to review; this with the caveat that, for a given site, a list might not be formed at all. There are numerous reasons why a site might not qualify for even a singular ERA task. Where these come to bear, the ecological treatment is (or should be) in the written form only, with no more than a paragraph or two explaining why an assessment was not needed. The responsible ERA practitioner understands the value of such brief narratives; they preclude HQ computation, something that should be avoided at all costs if possible. (Note that none of the studies of this compendium call for standard HQ computation. At best, they ask interested parties to prove how erroneous HQs can be, and to demonstrate that HQs are of no utility.)

As to why a site might not have a valid list of organisms to stand for assessment, there are also numerous reasons. As part of a brief, categorical review of the book's studies, we first see that a fair number of them are directed at the discovery of reasons for sites being without receptors worthy of assessment. Interested parties, then, are asked to investigate: if chemical exposure actually occurs (e.g., Studies #5 and #33), if enough animals of any one type occupy a contaminated space to matter (e.g., Study #11), and if a certain receptor group is present altogether (e.g., Study #29). An overarching tasking for interested parties calls for compiling an evidentiary record of instances where site species have been impacted (Study #1). Assuming a site does have at least one valid-for-assessment species, other studies in the compendium next ask interested parties to validate available screening and other assessment tools for their helpfulness (e.g., Studies #14 and #25), and to substantiate a need for benchmarks altogether (e.g., Study #2).

Where available tests and other tools are found to be wanting, still other studies encourage parties to devise serviceable databases that relate chemical body burdens to health effects (e.g., Studies #16 and #21).

In recognition of the reality that receptor exposures to site contamination are anything but new, still another category of study encourages parties to conduct direct health status assessments in the field (e.g., Studies #9 and #18). A related study category solicits the dosing of animals in a non-traditional way, where the natural chemical weathering that occurs with time is physically incorporated into the experimentation (i.e., Study #19). In a progression of approaches to understand if harmful effects from chemical exposure can arise, enthusiastic individuals are called upon to modify the conventional animal-dosing design (Study #20), and to dose terrestrial environments so as to create opportunities to monitor ecological effects, should they arise, over an expanded timeline, (i.e., Study #12). Perhaps the most ambitious study (Study #34) directs parties to probe the psyche of those who apply the ERA process religiously, and to a point where a certain fatal flaw—that ERA practitioners avow an unwillingness to ever deviate from its design—might be exposed.

The reader should not think that the 30-odd studies of this compendium exhaust the possibilities of what we need to still learn in order to move ERA along. The studies probe different areas of ERA, and do so from non-traditional angles. If nothing else, the diversity of research topics the studies embrace can and should serve as a springboard for the enthused ERA practitioner to explore still other facets of ERA. If an element of ERA doctrine doesn't well resonate with you, do not doubt your perception. Instead have the element investigated or taken to task as the case may be. It is suggested to procedurally first bin the element, casting it as either a task or a step that seems to not make sense or be necessary, or as a task or step that doesn't work (e.g., with it providing an inaccurate measure). Then, develop a study design to resolve the necessity for the questionable task or step (addressing the first case), or to illustrate that the said task or step either does or does not result in inaccuracy (addressing the second case).

The two situations profiled above suggest several worthwhile studies. With regard to the first one, the premature and ill-fated aspects of the PCL exercise aside, deficiencies in our toxicity databases are nevertheless noted through it. Setting aside, too, the inappropriateness of PCL development for a mouse and a shrew (for cleanups do not proceed for the protection of these species), a detail not mentioned above invites certain study. There were no PCLs developed for these two species in the case of the metal, mercury. Evidently, this happenstance traces back to the non-availability of a mammalian TRV for mercury, regardless of the metal's chemical form; ERA practitioners are surely fluent with assessments where the contractor has grabbed at the lone TRV he/she could locate, fully aware that it was for a chemical form different from that occurring at a site. There is still more to put aside with the trouble-ridden PCL exercise, as in mammal toxicity having played no role in PCL development. (Perhaps the two avians of the exercise, the robin and the bobwhite, can carry the way in the case of mercury, but lest this author be too accommodating, the reader is reminded that there was no legitimate trigger for the PCL exercise in the first place.) While HQs may be the ruination of ERA, and while this book certainly does not condone their computation at any point in ERA investigations, we nevertheless find ourselves duly reminded that the field is without an ingestion-based toxicity expression for mammals, for mercury. At the very least, then, it is suggested that a NOAEL and LOAEL, or perhaps some other more relevant toxicity expression should be developed for this metal.

Interestingly for the same PCL exercise, there was a second case of a lacking TRV, this one pre-empting calculations for the two avian species for their site exposures to antimony. Other than marveling at the absence of such toxicity information at this late date, practitioners should realize the opportunity to better ERA's lot in this area. Lest we conclude that the data gap can only be filled in the traditional/linear-thinking fashion (i.e., by dosing birds to produce a workable TRV), there could first be exploration into the validity of the need. Perhaps, then, a desktop investigation is primarily what is needed. It could review the frequency with which antimony presents as a "player" at sites, both in an overall sense, and also as it specifically relates to birds. Perchance, antimony tends to often co-occur with other contaminants. What might distill from such a review, and before resorting to the syringe or gastric intubation apparatus, is that a lacking antimony TRV for birds is not at all critical. Inquisitive individuals might next want to investigate why there is no move afoot to better the TRVs that we do have. Why should a TRV derived from a mouse study that was conducted two or more decades before ERA came into being serve in perpetuity (i.e., to never be re-examined or possibly bested)? The reader should appreciate that the list of worthwhile investigations to pursue, all growing out of an initial one, are limited only to one's thinking.

With regard to the lackluster 13-acre site, there is also new ground to encroach upon for ERA betterment or, dare it be said, abolition. While several studies (#s 4, 26, 27) are oriented to arriving at the parameters that identify contaminated sites to legitimately submit to assessment, a holistic approach to subject matter review could indicate that ERA, in the context we've been discussing all along, is fully purposeless. Perhaps our highlighted 13-acre site truly does not support animals, and it is for that reason that no ERA tasks were suggested. Perhaps, too, sites that are fully populated do not need to submit to ERA either; how bothersome could a site's chemicals be to receptors, if receptors, in appropriate numbers, are regularly observed occupying the site?

Should a fair amount of the described studies come to fruition, it is the author's anticipation that certain common denominators will show through. We may learn that site contamination is not harmful to biota, that we are incapable of assigning risks, that we have no need to try and assign risks, and—hold onto your seat—that there is no need for ERA altogether, the last of these undoubtedly a bitter pill to swallow for many. Dramatizing this last point would serve a beneficial purpose. It is a reality that people become unwell in some way. They meet with a health care professional or get admitted to a hospital. Some form of medical care is rendered, and in many cases the affliction is lessened if not fully cured. The field of medicine, for diagnosis, treatment, and prognosis exists. People seek to geographically relocate, be it on a temporary basis as in commuting to work daily, taking a summer vacation, or settling into a new community. People reach their destinations via automobiles, buses, trains, and planes. Transportation exists. People want the entry points to their homes to be safeguarded against intruders and to provide overall protection. They secure the services of locksmiths who install the necessary appurtenances. Locksmithing and home security exists. Chemical releases occur in our world, sometimes through deliberate means and sometimes through accidental means. Environmental media bear chemical loads and the non-human species that interact with the media stand to potentially be impacted. There is an interest in determining the probabilities of chemically exposed non-human species developing health effects or dying off outright. Unlike the preceding examples, we are unable to deliver; we cannot determine the probabilities. Ecological risk assessment does not exist. (It is an ultimate irony—the field is called ecological risk assessment, but a way does not exist to assess ecological risk. This author finds it incredulous that he is employed in a field that does not live up to its name. He finds it incredulous, as well, that so many involved in the "field" are fully unaware of the field's absence of ability.) Determining, in a laboratory setting, the chemical concentration in an artificially amended medium that is harmful (perhaps deadly) to commercially bred test organisms, and then comparing that concentration to that measured in the field, does not amount to ecological risk assessment. And why would someone want to know of the probability of biota becoming harmed when site biota have not succumbed over multiple decades? For the contaminated environment context that this book addresses, the odds favor the discovery that there is no need for ERA altogether.

We return to an ecological setting and a phenomenon relevant to several of the suggested studies. Sediment samples of a contaminated waterbody's bioactive zone and that of the same waterbody's upstream reference point are collected. Benthic macroinvertebrates are removed from the samples and painstakingly keyed out to the genus level in support of respective community characterizations. A striking difference in macroinvertebrate benthic assemblages is observed. The contaminated portion has 7 species, six of which are recognized pollution-tolerant ones. The reference location has 11 benthic macroinvertebrate species, 9 of which are recognized pollution intolerants. The data tell a story, and it is embellished when sediment sampling is subsequently conducted less than a quarter-mile downstream of the contaminated waterbody region. There, the assemblage almost perfectly mirrors that of the upstream reference location. Some would want to make a great issue of the arrangement at the waterbody's contaminated portion, especially when it's so very clear that the chemical(s) at play are at cause. Does the community assemblage difference matter, though? Who is it bothering? Why would anyone find themselves bothered? Consider that the waterbody has borne its contamination for several decades, and that the assemblage shift is not by any means something new. To those who would suggest that fish are affected, is there information to support such a contention? If not, how could fish be assessed—without, of course, resorting to working with caged fish? Would anyone even know of the admittedly chemical contamination-caused assemblage shift had the sediment samples not been collected? Is the waterbody any less productive over its contaminated portion relative to other upstream and downstream portions? How so? Is the contaminated portion biologically dead? Are the waters there anoxic? What isn't the contaminated waterbody portion doing that it should be?

Overwhelmingly, the record shows that ecological receptors and the ecosystems present at contaminated sites are free of impact. Despite this, the ERA community remains focused on trying to uncover differences in measure for the living forms at sites that trace to unleashed chemical contamination. For all of their efforts, they've very little to show. It would seem that it's time for ERA to stop trying to demonstrate that something is wrong, and to be willing to hear that biota living with contamination could be just fine; hence this compendium of studies. They are designed to allow for the truth to be told. If we are to discover that animals do not contact contaminated sites sufficiently to be harmed by them, so be it. By way of example, if non-manipulated frog embryos prove to be far less sensitive to water column chemicals than are artificially manipulated embryos, we can acknowledge the bias of an indoctrinated standardized test that says otherwise. And so on.

It is the author's hope that efforts expended in conducting the suggested studies will lead to greater clarity for ERA.

Index